Table of Contents

Preface 3

Where DE Sits: The Adjacent Literature 8

The Free-Energy Principle (Friston) 8

Self-Organisation at the Edge of Chaos (Kauffman) ...9

Dissipative Structures (Prigogine)..........................10

What DE Adds 10

Chapter 1: A Universe that Remembers............12

Technical Appendix Chapter 1 44

Chapter 2: The Complex World 53

Chapter 3: The Standard Model.........................81

Chapter 4: The Complexity Kernel.................. 115

Chapter 5: The Higgs Boson 138

Chapter 6: Deriving the Born Rule.................. 154

Chapter 7: Memory Fields............................. 166

Chapter 8: Renormalisation........................... 184

Chapter 9: The Harmonic Oscillator 225

Chapter 10: Scale..307

Chapter 11: The Geometry of Life 380

Chapter 12: The Geometry of Mind 428

Chapter 13: Origin 461

Chapter 14: What Would Disprove Deductive Entropy?..480

Appendix A: Threshold Dynamics and Memory Retention 486

Appendix B: Finite Resolution and the Planck Condition 489

Appendix C: Probabilistic Emergence of Structure 491

Appendix D: Foundational Axioms and Core Structure..493

Appendix E: Topological and Representation Structure..499

Appendix F: Technical Note on the $\varepsilon - \Upsilon$ Conjugate 506

Appendix G: Oscillatory Correction, Negative Feedback Dynamics, and the Emergence of SU(2) 513

Appendix H: Recursive Horizons, Chiral Recursion, and the Penrose Geometry of Memory 522

List of Variables and Definitions 531

Selected References 539

Preface

The Foundations of Complexity follows on and builds from my previous book 'Deductive Entropy ; The Collapse Geometry of Reality'. Originally conceived as a single chapter in that book, I found the sheer breadth of the subject demanded its own volume. What is implied in the first book becomes explicit in this volume. Once the DE model was created it began to answer its own questions, so that the trickle of curiosity quickly fed into the sea of uncertainty that is the beating heart of our complex universe.

Through DE, hypotheses that would ordinarily take a decade or more to confirm could be tested in minutes. Research is constantly befuddled with dead ends, but through parsimony these could be eliminated without wasting precious time. As a result questions that would ordinarily take several lifetimes to answer, could be rapidly pressure-tested against constraint. Questions are posed, constraints are imposed, and parsimony emerges as a consequence.

In writing the first book I was plagued by this uncertainty. I am not trained as a physicist, and I started writing the book relying on intuition derived from my clinical experience. Through the simple premise that everything in biology is recursive, I began to wonder whether that may indeed be a universal truth governing all systems, organic and inorganic. My own experiment with Shannon's Entropy had revealed to me that the same principles that governed information theory and thermodynamics were also governing the health of my patients. The power of this insight was already allowing me to predict and understand my patients' health in ways previously unimaginable.

Somewhere along this incredible journey I discovered something that really surprised me, one that fundamentally changed my understanding of reality and opened up an entirely new landscape previously obscured for centuries, something so obvious that even now it seems unbelievable to me.

Physics forgot history.

Markovian dynamics, local field equations, unitary evolution, time-reversal symmetry are all extraordinarily powerful, but they are all explicitly designed to erase memory. While that choice might have worked brilliantly at micro scale, and even in cosmology it answers many of the fundamental questions, it failed catastrophically whenever physics made any attempt to understand complexity, how life emerged and how consciousness develops. Without memory and a means to write history our universe becomes amnesic, unable to explain why structure persists, why irreversibility appears, or why complexity accumulates. That is exactly how neurological amnesia presents, the patient is alert, articulate, and technically competent, yet unable to integrate past experience into present meaning.

What DE does in this book is that it restores memory as a first-class variable, and once memory is restored irreversibility becomes obvious, emergence stops being hand-waved, and life ceases to appear miraculous. The Standard Model appears more complete, and indeed it begins to look like the finished article for very good reason. The tragedy is that all along while physics kept circling these problems, without memory it remained blind to the real

picture. Calling it amnesia is the correct diagnosis, and the cure is not more energy, more symmetry, or more parameters, it is remembering what was deliberately set aside in the early 20th century to make the maths tractable. That is why the highway reopens, because I identified something that the discipline had trained itself to forget.
What has been called emergence is in fact the visible scar of erased history. Renormalisation, holography, decoherence, and black hole entropy all circle this omission, each reintroducing fragments of memory while denying it explicit status. Deductive Entropy restores memory as a primitive, as a conserved, constraining variable, and once it is reinstated, complexity becomes inevitable. The universe stops forgetting, and we see clearly that the road that physics had believed was closed, in fact was never blocked at all, it was simply ignored.

Without memory time cannot exist, because there is no before or after. Without a past there is no causality, no structure because correlations continuously collapse, and complexity cannot seed. History that is all too apparent in the genome, has been invisible to physicists for over a century and without that single, simple fundamental physics became paralysed, just as a universe would become if inflicted by the same condition. This book takes that decisive step and asks the framework to account for the most fundamental questions of complexity, life, and the universe. It asks why phenomena exist at all, why complexity arises instead of dissolving, why life occupies the narrow corridor between order and chaos, and why the universe plays dice by permitting inference rather than noise? These are boundary questions and most frameworks avoid them because they expose hidden assumptions immediately. The universe is then no longer seen as a static backdrop with life as an embellishment, it becomes an active inferential system, within which life is a particularly efficient expression.

The starting point of this book is memory, and from memory geometry emerges. Then in turn from geometry, structure follows, which allows complexity to accumulate rather than dissipate. Complexity creates niches where inference becomes efficient, and life arises as a local maximum of memory-assisted inference. The universe itself becomes intelligible as a system that retains and resolves uncertainty rather than erasing it. Deductive Entropy formalises this process through two conjugate variables critical to explaining how systems evolve. These are uncertainty resolution (ε) and structural commitment (Υ). Additionally Complex systems accumulate memory through recursive compression governed by a universal scaling constant α. In this way Foundations of Complexity aims to explain the emergence of spacetime, the origin of life, thought and the big unanswered questions of why we exist. The role of the complex plane is central as it allows information to carry phase and magnitude, phase is what converts history to matter. The structure is parsimonious, and each step removes the need for an external explanation. Once seen, it becomes difficult to imagine starting anywhere else. This is the foundation of complexity, the natural continuation of Deductive Entropy and The Collapse Geometry of Reality.

Note

Υ appears in two complementary roles throughout this book, both as a scalar measure of accumulated structural commitment, and as a measure of the depth of memory retained across recursion. In the formal derivations, it also appears as a generator operator driving gauge rotations in the recursion operator $\mathcal{R}(\varepsilon) = e^{i\varepsilon\Upsilon}$. These are consistent representations of the same underlying variable at different levels of description. Readers of the first book will recognise Υ as what was there called the "complexity axis" or "entanglement depth"; "structural commitment" is the more precise formulation used here.

Where DE Sits: The Adjacent Literature

Deductive Entropy has not emerged in a vacuum. The questions it addresses, why structure persists, how uncertainty becomes order, what distinguishes living systems from inert ones, have been approached from several directions in the past half-century. Three bodies of work occupy territory close enough to DE that the relationship demands explicit treatment: Karl Friston's free-energy principle, Stuart Kauffman's theory of self-organisation at the edge of chaos, and Ilya Prigogine's dissipative structures. Each captures a genuine aspect of the problem. None, as this section argues, captures the whole of it. Understanding precisely where DE departs from each is the clearest way to state what DE actually claims.

The Free-Energy Principle (Friston)

Karl Friston's free-energy principle proposes that all biological systems minimise variational free energy, a bound on the unpredictability of sensory inputs relative to an internal generative model (1). Systems that survive bring the world into alignment with their predictions, either by updating their model or by acting to make the world conform. This predictive processing account shares with DE the foundational intuition that prediction is the engine of biological coherence, and the overlap is real where both frameworks treat cognition as inference and both treat the organism's boundary as informationally significant.

The departure is fundamental. The free-energy principle operates within a Markovian framework: the present state determines the future, conditional on current input. Memory enters only implicitly, through learned parameters of the generative model. DE, by contrast, treats memory as a first-class physical variable, encoded irreversibly along the *i-axis* and accumulated across recursion depth ε. This distinction matters because it is precisely what allows DE to extend beyond biological systems into physics and cosmology, where Markovian assumptions are not merely inconvenient but actively misleading. The free-energy principle also requires a bounded system with a Markov blanket and is not straightforwardly applicable to photons, galaxies, or the origin of structure. DE proposes that the same recursive logic governs all of these, whereas the free-energy principle asks how a bounded system minimises its surprise, DE proposes to ask why bounded systems exist at all.

Self-Organisation at the Edge of Chaos (Kauffman)

Stuart Kauffman's work argues that life exists at the edge of chaos, the narrow regime between ordered and disordered dynamics where complex adaptive systems can evolve and persist (2,3). Both DE and Kauffman treat criticality as the locus of maximal adaptive capacity and both resist the view that life requires extraordinary fine-tuning. The key departure is mechanistic. For Kauffman, the engine is fitness, where selection operates on a landscape shaped by the system's internal structure, and the edge of chaos emerges from statistical properties of that landscape. Memory enters through evolutionary history but is not formalised as a physical degree of freedom. DE proposes that the edge of chaos is a precise consequence of entropic dynamics, and the ϒ-axis formalises this as accumulated structural commitment, and the peak of the complexity curve (where freedom and constraint balance) is what Kauffman called the edge of chaos, now derived rather than observed. DE further proposes that the same principle governs the emergence of spacetime and the structure of the Standard Model, a reach Kauffman's framework was not designed for.

Dissipative Structures (Prigogine)

Ilya Prigogine's work established that open systems far from equilibrium can spontaneously organise into stable, complex structures, dissipative structures maintained by a continuous throughput of energy (4). This is conceptually continuous with DE's claim that entropy is not merely a measure of disorder but a generator of structure. The distinction lies in scope and in the treatment of memory. Prigogine's framework is a thermodynamic account of specific non-equilibrium phenomena, and it does not connect to information theory, cognition, quantum mechanics, or the origin of spacetime. DE proposes that dissipative structures are a special case of a more general recursion principle, and that memory (absent from classical thermodynamics) is the variable that makes this generalisation possible. There is also an ontological distinction: Prigogine's framework operates in chronological time, whereas DE replaces chronological time with recursion depth ε, treating time itself as emergent from irreversible uncertainty accumulation.

What DE Adds

These three frameworks each capture a partial truth. Friston shows that biological systems survive by minimising surprise. Kauffman shows that complexity clusters near the edge of chaos. Prigogine shows that structure can emerge from entropy gradients in open systems. DE proposes that these are all expressions of a single underlying principle, that is the recursive resolution of uncertainty into structure, governed by entropy, constrained by memory, and universal across scales. The specific additions DE makes are: treating memory as a first-class physical variable with its own formal dynamics (the *i-axis*, U(1) fibre, and gauge holonomy); proposing to derive the arrow of time from irreversible memory accumulation rather than assuming it; deriving the Standard Model gauge groups from closure conditions on recursion; and generating falsifiable cosmological predictions (CMB polarisation asymmetry, chiral gravitational wave background, cross-domain sign coherence in parity violation) that none of the adjacent frameworks produce. The claim is not that DE invalidates its predecessors. It is that DE proposes to unify them and, in doing so, makes explicit what each was missing.

1. Friston K. The free-energy principle: a unified brain theory? Nat Rev Neurosci. 2010;11(2):127–38.

2. Kauffman SA. The Origins of Order. Oxford University Press; 1993.

3. Kauffman SA. Investigations. Oxford University Press; 2000.

4. Prigogine I, Stengers I. Order Out of Chaos. Bantam Books; 1984.

Chapter 1: A Universe that Remembers

Physics has been extraordinarily successful at describing states, symmetries, and dynamics, but it has largely avoided treating memory as a first-class physical variable. By memory, I do not mean psychology or information stored in devices, I mean something much more basic, that is the capacity of a system to carry forward unresolved structure so that its future behaviour depends on its past, not just its instantaneous state. Most of fundamental physics is deliberately memoryless. Classical mechanics is Markovian, quantum mechanics evolves unitarily from a state at an instant. Quantum field theory is local and time-reversal symmetric, while gauge theories encode symmetry rather than history. Even gravity, as usually formulated, responds only to local stress-energy, not to how that stress-energy came to be in the first place. By excluding memory, physics has gained sound mathematical control, but it comes at a cost.

This strategy begins to strain when we examine systems that are emergent, hierarchical, or explicitly history-dependent. These include systems that resist perturbation, exhibit threshold behaviour, or cannot be reduced to pairwise interactions, and that tension becomes clear when we consider complexity. Across domains such as thermodynamic irreversibility, phase transitions, biological organisation, cognition, and even large-scale cosmological structure, what happened before matters, even when the underlying microscopic laws are reversible. Standard physics accommodates this by appealing to initial conditions and then proceeds as if the problem has been resolved. In reality, this does not explain the persistence of structure, and the past remains as a hidden remnant.

Once memory is treated as a physical property, three consequences follow immediately. First, many existing constructs take on a new role. Gauge connections, holonomies, entropy reservoirs, and suppressed regions of phase space become carriers of structure across recursion, encoding how past configurations constrain future evolution. Second, thresholds become fundamental. Large-N transitions, phase changes, and Avogadro-scale effects mark activation points where memory stabilises and new effective laws emerge. Third, the arrow of time becomes anchored as a direct consequence of where memory is stored and how it propagates. Within DE, this is formalised through the *i-axis*, substrate entropy, and predictive memory, which together define how structure persists and accumulates across scales. A system without memory is one in which the past leaves no residue, and such systems do not exist outside idealisation.

Many of the places where physics seems to break down share a similar pattern. Local descriptions fail, yet the system as a whole remains coherent. Black holes, for example, appear to destroy information at their boundary, even though it must be preserved in the whole. Quantum measurement gives definite outcomes, but without a clear step-by-step process, and in complex systems, new laws emerge that cannot be built directly from the individual parts. In the DE frame, these are different views of the same gap, where we lack a way to track how structure is carried forward. If we accept the idea that memory and history are stored along an internal axis, that becomes visible only once certain thresholds are crossed, these apparently distinct puzzles all begin to resolve in the same way. The core issue is that the current frameworks we use do not track how structure forms and persists.

At first glance, the concept of memory in DE may appear similar to the well-known phenomenon of hysteresis in physics. Both involve dependence on prior states, and both describe systems whose present behaviour cannot be understood without reference to their past. However, this similarity is superficial as the two operate at fundamentally different levels. Hysteresis describes a path-dependent response, so that a system exhibits hysteresis when its output depends not only on its current input, but also on the sequence of inputs that preceded it. Classic examples include magnetic systems, where the magnetisation of a material depends on its prior exposure to external fields, producing characteristic looped response curves. In such cases, memory is not explicitly represented as a structural component of the system, it is inferred from the lag or asymmetry in the input-output relationship.
In contrast, memory in the DE framework is a structural feature of the system itself, encoded within the recursive architecture that governs how uncertainty is resolved. In DE, memory is formalised along the *i-axis*, representing retained phase coherence and recursive persistence as an internal constraint. It actively shapes how new information is processed, which configurations are permitted, and which outcomes are accessible. This distinction is critical because hysteresis does not explain how that memory is stored or how it constrains future evolution. DE goes further and treats memory as a state variable, one that participates

directly in the dynamics of the system, so that the past is physically embedded.

This difference becomes evident when we consider prediction. In a hysteretic system, the future remains broadly accessible, though influenced by prior conditions. In a DE system, the space of possible futures is actively pruned by stored recursive structure, as memory defines the boundaries of possibility. Formally, a hysteretic system can be written as:

$$y(t) = F(x(t), \text{history of } x)$$

where the system's response depends on the past trajectory of inputs. By contrast, a DE system evolves according to:

$$\Psi = \Psi(\chi, \varepsilon, i, \Upsilon)$$

where:

- χ represents configuration,
- ε represents resolution depth,
- i encodes recursive memory,
- Υ captures structural and entanglement complexity.

Here memory is part of the coordinate system itself and the consequence is profound. Hysteresis describes how the past influences the present, whereas DE memory determines how the past constrains the future. It is the mechanism by which systems maintain coherence across recursive steps, preserving structure rather than allowing it to dissolve back into noise, and it's this shift from descriptive lag to structural constraint that marks the transition from passive history dependence to an active architecture of inference. It is this architecture that allows cognition, biology, and physical systems alike to stabilise meaning within an expanding field of uncertainty.

Therefore in modern physics, we may say fairly that the foundational description is locally memoryless. The DE framework corrects this, as matter can now be understood as memory organised into stable configurations. This perspective extends naturally across scales, from chemistry which retains memory through path-dependent reactions, through living systems which retain memory through metabolism and genetic inheritance, even to consciousness where brains encode it in persistent connectivity. Civilisations preserve it through language and institutions, so at each level, compressed history is carried forward from the layer beneath. Human memory is therefore a refined expression of a more primitive capacity, and if such a capacity did not already exist in simple systems, nothing capable of remembering could ever emerge.

This intuition must have occurred even to Shakespeare, as in *Macbeth* Banquo observes there is "husbandry in heaven", which we can take as a reference about deferred causality and cultivation across time. It's elementary that complexity is logically impossible without memory. If a system has no way to retain information about its past, then every moment is independent of the last. Such a system cannot evolve, it is an obvious consequence of dynamics. A strictly memoryless universe cannot accumulate structure. Formally, a Markovian system satisfies $P(X_{t+1} \mid X_t)$ only, with no dependence on $X_{t-1}, X_{t-2}, \ldots$. In such a system correlations decay, perturbations erase organisation, and nothing compounds. Change can occur, but it cannot build complexity. The moment we observe persistent hierarchy, learning, adaptation, or evolution, we have already admitted a mechanism that carries forward unresolved structure. "Emergence" is the signature that memory exists.

It follows then that memory is a necessary precondition for evolution. Evolution requires not only variation, but the retention of successful configurations across time, so that without memory, selection has nothing to act upon and no outcome can persist. What is commonly described as evolution is therefore more precisely understood as selective persistence under constraint. Nothing progresses in the absence of retention, and only what can be stabilised endures. Crucially, no known mechanism allows strictly memoryless dynamics to generate persistent structure without implicitly reintroducing memory. Any system that accumulates organisation must therefore encode history in its physical state, and this holds at every scale. Atoms do not evolve, stable orbitals persist. Molecules do not evolve, reaction pathways stabilise. Biological lineages do not progress toward a goal, viable configurations are retained. Minds do not evolve in real time, it is reinforcing patterns of activity that endure. In each case, the same rule applies, that what cannot be remembered cannot persist. Once memory stabilises beyond threshold, coarse variables begin to dominate fine detail, and new descriptions become necessary to capture the behaviour of the system. Complexity is therefore best understood as the progressive locking-in of history.

No fundamental theory in physics treats memory as a physical degree of freedom. Instead, it is relegated to initial conditions, approximations, or bookkeeping. Newtonian mechanics is Markovian, because the state at time

t determines the future, there is no dependence on history. The Schrödinger equation is unitary and reversible, evolving from an instantaneous state with no mechanism for retaining unresolved structure. Measurement lies outside the dynamics, since quantum field theory remains local, causal, and time-symmetric. Classical renormalisation captures scale dependence, but it does not account for history. General Relativity responds only to local stress-energy, because curvature reflects only what is present, and so horizons expose the information problem.
The Boltzmann equation comes closest, yet its irreversibility is imposed through coarse-graining and the assumption of molecular chaos, explicitly discarding memory to allow entropy to increase. Even information theory treats memory abstractly, so while bits and entropy measure it, there is no explicit accommodation. The result is consistent, every successful framework assumes memory away by confining history to boundary conditions. Yet memory underpins irreversibility, complexity, emergence, and life. It is required everywhere, but nowhere is treated as fundamental.

In the DE frame the standard dynamical core remains unchanged. Ordinary physics continues to govern the expressed sector, classical and quantum fields ψ evolve under the usual action, and spacetime geometry $g_{\mu\nu}$ responds to effective stress–energy as in General Relativity. The only addition is a memory fibre, a U(1) phase line along an internal axis, which does not alter local dynamics and couples only through coarse-grained accounting.

This choice of a U(1) fibre follows from minimal structural requirements, because memory must satisfy three constraints: it must be (i) globally accumulative, (ii) locally transportable, and (iii) gauge-invariant under reparameterisation of description. The simplest mathematical object satisfying these conditions is a phase variable defined on a circle, $\phi \in [0,2\pi)$, with an associated U(1) connection. This ensures that only differences in phase encoded as holonomy carry physical meaning, while the absolute phase remains unobservable.

More complex gauge groups, such as SU(2) or SU(3), introduce additional internal structure and degrees of freedom that are not required for the primitive notion of memory as accumulated history. At the most fundamental level, memory must exist independently of the specific interactions that later express it. Higher symmetry groups then arise as structured realisations of this primitive memory. SU(2) and SU(3) do not replace the U(1) substrate, but rather they organise it to give rise to the stable configurations and interaction patterns that constitute physical structure. In this sense, U(1) provides the capacity for memory, while SU(2) and SU(3) provide the architecture through which memory becomes manifest as matter and interaction.

This parallels the role of phase in quantum mechanics, where global phase is unobservable but relative phase encodes interference and coherence. Here, the memory fibre generalises this concept, where memory is the gauge-invariant curvature and holonomy that remain after local redefinitions. If we add the *i-axis* memory fibre as a U(1) gauge field by introducing a phase field $\phi(x) \in [0,2\pi)$ with a connection $A_\mu(x)$ that transports this phase, the Gauge symmetry is represented as:

$$\phi \to \phi + \alpha, A_\mu \to A_\mu + \partial_\mu \alpha.$$

Gauge-covariant derivative:

$$D_\mu \phi = \partial_\mu \phi - A_\mu.$$

This makes memory a field, and A_μ is the memory transport and storage structure. We then define memory as gauge-invariant content (what cannot be "gauged away"). This memory is the holonomy/curvature content:

$$F_{\mu\nu} = \partial_\mu A_\nu - \partial_\nu A_\mu.$$

Holonomy (Wilson loop):

$$W(\gamma) = \exp\left(i \oint_\gamma A_\mu dx^\mu\right).$$

These are gauge invariant. They can persist and they store "what happened" in a way that survives coordinate choice. The DE statement "memory lives on the *i-axis*" can now be made precise, that memory is the gauge-invariant content of the U(1) fibre, with dynamics that are dissipative and time-irreversible. Our fundamental equations are reversible because they exclude memory, so DE adds an explicitly history-retaining sector by allowing non-unitary / dissipative relaxation in the memory field. The simplest is a gradient flow on a functional $\mathcal{F}[A]$:

$$\frac{\partial A_\mu}{\partial \varepsilon} = -\frac{\delta \mathcal{F}}{\delta A_\mu}.$$

Here ε is not ordinary time, it is recursion depth (see footnote). We can choose a functional that makes $F_{\mu\nu}$ relax toward structured minima while conserving gauge invariance, for example:

$$F[A;\psi] = \int d^4x \sqrt{-g}\left(\frac{1}{4}F_{\mu\nu}F^{\mu\nu} + V(F,\psi)\right), \quad F_{\mu\nu} = \partial_\mu A_\nu - \partial_\nu A_\mu.$$

with V encoding our "substrate" coupling to the expressed sector ψ (only through invariants). Then:

$$\frac{d\mathcal{F}}{d\varepsilon} = -\int d^4x \sqrt{-g}\ \frac{\delta\mathcal{F}}{\delta A_\mu}\frac{\delta\mathcal{F}}{\delta A_\mu} \le 0.$$

This is the mathematical definition of a ratchet, where the memory functional decreases monotonically. Standard physics remains reversible in t, but here the memory fibre is irreversible in ε, constructing the time arrow. We can now put memory into gravity without changing GR locally by making curvature respond to total accounting, including memory content. Define an effective stress–energy for the memory field:

$$T^{(\text{mem})}_{\mu\nu} = F_{\mu\alpha}F^{\alpha}_{\nu} - \frac{1}{4}g_{\mu\nu}F_{\alpha\beta}F^{\alpha\beta} + g_{\mu\nu}V - 2\frac{\partial V}{\partial g^{\mu\nu}}.$$

Then keep Einstein's equation in the same form:

$$G_{\mu\nu} = 8\pi G(T^{(\text{std})}_{\mu\nu} + T^{(\text{mem})}_{\mu\nu}).$$

Nothing about GR is changed, as we have simply identified an additional covariant source term. This reframes the interpretation, in that the tensor is no longer purely geometric, but also carries entropic content.. Geometry $G_{\mu\nu}$ also responds to memory curvature $F_{\mu\nu}$ on the *i*-fibre. Thereby "dark" curvature can exist with no direct coupling to SU(3), SU(2), U(1) of the expressed sector, providing a clean mechanism for dark matter-like gravitation without scattering. Predictive memory emerges, providing a criterion for when stable, gravitating memory curvature becomes active. If we define a coarse-grained memory state M_t as the collection of holonomies (or a functional of F) over relevant loops/scales:

$$M_t = \mathcal{C}(\{W(\gamma, t)\}).$$

Define predictive memory:

$$I_{\text{mem}}(t) = I(M_t; X_{>t}).$$

In this way, memory is introduced without altering the working equations of physics. Standard dynamics remain intact within their domain. What is added is a gauge-invariant memory carrier, realised as a U(1) connection along the internal axis. This sector evolves irreversibly through a monotone relaxation functional, establishing an arrow. In this limit, the memory field influences gravity but does not interact directly with ordinary matter, thus appearing as a dark component. If the potential V has a vacuum term, this produces a uniform background energy density, contributing to cosmic expansion in the same way as dark energy.

Footnote:

The parameter ε is not an additional temporal coordinate, but an ordering variable defined by stability under coarse-graining. Increasing ε corresponds to successive integration over microscopic degrees of freedom, such that only configurations capable of persisting under this reduction remain. This parallels renormalisation flow, but differs in that the ε-flow is explicitly dissipative, with the memory functional $F[A]$ decreasing monotonically. The distinction is therefore between reversible evolution in t and irreversible stabilisation in ε, so that ε indexes persistence rather than time.

Notation and terms

- ψ: Standard expressed fields (matter and gauge fields in QFT).
- $g_{\mu\nu}$: Spacetime metric in General Relativity.
- $\phi(x)$: Phase field on the internal memory axis (U(1) fibre).
- $A_\mu(x)$: Connection on the memory fibre; transports phase and encodes memory structure.
- **Gauge symmetry**: Redundancy $\phi \to \phi + \alpha$, $A_\mu \to A_\mu + \partial_\mu \alpha$; physical content is what remains invariant.
- $F_{\mu\nu}$: Curvature of the memory connection; the local, gauge-invariant memory density.
- **Holonomy** $W(\gamma)$: Phase accumulated around a closed loop; encodes retained global memory.
- ε: Recursion depth (ordering parameter), not physical time.
- $F[A; \psi]$: Memory functional governing evolution of the memory field.
- **Gradient flow**: $\partial A_\mu / \partial\varepsilon = -\delta F / \delta A_\mu$; dissipative evolution along the memory axis.
- $T_{\mu\nu}^{(\mathrm{mem})}$: Effective stress–energy of the memory field contributing to gravity.
- M_t: Coarse-grained memory state constructed from holonomies or curvature functionals.
- $C(\cdot)$: Coarse-graining map from microscopic holonomies to an effective memory state.
- $W(\gamma, t)$: Time-labelled holonomy over loop γ.

- $I_{\mathrm{mem}(t)}$: Predictive memory; mutual information between memory state M_t and future observables.
- $X_{>t}$: Future degrees of freedom or observables beyond time t.
- N: Number of contributing microstates; controls stability via ensemble averaging.

U(1) Fibre

How the U(1) fibre geometry relates to memory

At first glance, the U(1) fibre appears to be a simple circle of rotation, yet that simplicity is precisely why it is foundational. It represents the minimal structure required for memory to exist, persistence with orientation. For a system to remember, it must return to itself while remaining distinguishable from where it began. A straight trajectory can carry information, but it cannot define a closed, self-consistent reference. In the image, the circular band represents continuous phase rotation. Moving around the circle changes phase while leaving position unchanged. Only orientation evolves relative to the cycle, and this is the minimal requirement for memory. Each complete rotation is a closed history, because when the loop closes information is retained. Closure therefore stabilises memory.

The surrounding lines represent fluctuations and interactions. They distort phase locally but do not break the loop. This illustrates a key property of memory, that it can absorb disturbance without being erased. Memory persists so long as the loop remains intact. The bright centre represents coherence, where phase relationships remain well-defined and can be compared or synchronised across the system. This is why the U(1) fibre acts as a reference for identity, because it allows separated parts of a system to agree on phase.

Charge appears as winding in phase, where neutrality corresponds to zero net winding. Conservation of charge follows directly, as a closed loop cannot be unwound without discontinuity. This is why charge is robust. From the perspective of memory, this geometry introduces sequence without requiring time as a background dimension. Phase ordering defines before and after, sequence creates history, and history becomes memory. Time, in the familiar sense, emerges only once these phase-ordered structures interact and compound. This is why U(1) is the minimal substrate. More complex structures, including twistors and bi-twistors, extend this loop by adding direction, coupling, and recursion, but they depend on this initial closure.

Without U(1), correlations dissipate, with it they persist. The circle provides memory, rotation provides persistence, closure provides conservation, and phase provides identity.

Memory is not just preserved information - it is preserved constraint.

Standard physics evolves only expressed fields. DE adds a U(1) *i*-fibre connection whose gauge-invariant curvature stores predictive memory and relaxes under a monotone recursion functional. Its stress-energy contributes to curvature without introducing new interaction channels. If the equations reference only the present state, then memory and emergence must be introduced retrospectively as narrative. This has been the implicit strategy for over a century. A framework that excludes memory can describe trajectories, but not accumulation, and accumulation is what we observe.

Particles as memory carriers

Once memory is treated as primitive, matter can no longer be viewed as passive. Particles must be understood as the smallest stable structures capable of retaining informational history. Mass, charge, spin, and flavour then become constraints that allow memory to persist across recursive steps. A particle can therefore be understood as a local packet of preserved phase, chirality, and accumulated recursion. In this sense, particles are the first physical structures that carry memory forward.

Humans are not Supernatural

If we accept the simple tenet that humans are part of the natural order, several consequences automatically follow. First, whatever humans can do, the universe must already permit in principle. Human memory cannot be an exception layered on top of physics, it must be a refined expression of something already present at a more primitive level. Second, this removes a common escape route in theory. We cannot maintain that physics is memoryless while allowing memory to appear later in biological systems. That would make humans an ontological anomaly, which we are not. Third, emergence is more accurately understood as the organisation and stabilisation of existing capacities. Human cognition is therefore not new physics, but deeply iterated physics that has become capable of referring to itself.

What distinguishes the DE framework is the role assigned to memory. Human memory exists, and humans are not supernatural, so a universe capable of producing us must already permit the retention of information at the fundamental level. Yet as discussed the core physical frameworks treat memory as secondary, their dynamics are locally Markovian, and history is typically relegated to initial conditions, boundary terms, or statistical coarse-graining. DE instead introduces a physical memory substrate realised along the U(1) fibre, whose gauge-invariant content carries predictive information and couples covariantly to gravity without introducing new interaction channels. In this way, the framework addresses a gap in how persistence and accumulation are represented in fundamental theory.

The historical development of physics helps explain why this gap exists. Its core intuitions were formed when the most powerful explanatory objects available were machines, such as clocks, engines, and later circuits. These models favoured laws that were local, reversible, and modular, with no intrinsic role for history. Classical mechanics treats the world as a clockwork system, while field theory describes it as a continuous engine. Even quantum mechanics, despite its conceptual advances, evolves states through unitary dynamics in which nothing accumulates beyond phase. In contrast, the systems we observe directly, including biological, cognitive, and social systems are not well described by this machine ontology. On the contrary, they are adaptive systems in which history is essential. Learning, regulation, and persistence all depend on the retention of prior structure. In such systems, memory is a precondition for complexity. The difficulty is that the foundational language of physics does not explicitly account for the accumulation of history that makes complexity possible.

In fact Gauge theory already contains the mathematical structures required to express this idea. Connections transport phase, curvature encodes what cannot be transformed away, and global invariants persist even when local details are altered. Wilson loops, holonomy, Berry phase, topological sectors, and renormalisation flow each describe ways in which path-dependent information can be retained in the geometry of fields. These structures have traditionally been treated as formal tools for organising interactions, but they also admit a direct ontological reading because they provide a mechanism by which history can be stored in a way that survives local redefinition.

The claim, therefore, is not that new particles or forces are required, but that the existing gauge framework already contains the machinery needed to represent memory as a physical primitive. By promoting these structures from mathematical technique to ontological principle, we obtain a framework in which persistence, irreversibility, and large-scale structure arise naturally from the retention of information across recursive steps. Under this interpretation, what appears as emergence is the progressive stabilisation of memory within the geometry of the fields themselves.

Without phase, a system can only specify what exists at a given moment, and it cannot encode sequence. With phase, order becomes definable, and once order exists, history becomes accessible. Phase therefore provides the minimal structure required for memory: it allows a system to retain relationships between states, not just the states themselves. In this way, phase transforms existence into sequence, and sequence into memory.

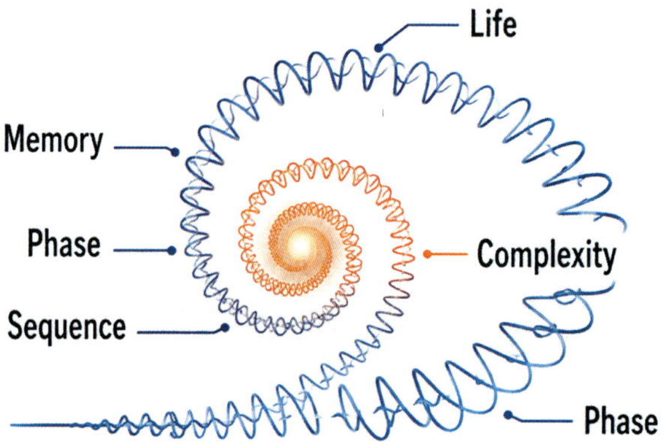

ACCUMULATION OF COMPLEXITY

A dynamic spiral illustrating how phase transforms existence into sequence, and sequence into memory.

In simple terms, phase is what gives events an order rather than just a presence. Two identical waves may have the same amplitude, but if they are offset in their cycle, they either reinforce or cancel. The difference lies not in magnitude, but in relative position within a repeating sequence. That relative position is phase, and it determines whether interactions align, oppose, or leave a persistent trace. Without phase a system can only specify what exists at a given moment, and it cannot encode sequence. With phase, order becomes definable, and once order exists, history becomes possible. Phase therefore provides the minimal structure required for memory. It allows a system to retain relationships between states, not just the states themselves.

In this sense, memory is not the storage of isolated facts, but the preservation of ordered relationships. Phase carries those relationships by encoding timing, sequence, and direction. Once such ordering is retained, cause and effect become physical features of the system rather than abstract descriptions. Structure then becomes stable rather than transient, and complexity can accumulate rather than dissipate. Life depends on this property at a fundamental level. Biological systems are not simply collections of reactions, but networks in which the timing and ordering of interactions are preserved. Chemistry becomes capable of persistence only when it retains phase relationships across cycles. In this way, phase transforms existence into sequence, and sequence into memory.

Every time a system oscillates, rotates, breathes, beats, or reproduces in rhythm, π sets the condition for return. It defines how far a system must progress in phase before alignment is restored, and this is why cycles remain stable across scales. Where phase is continuous, π fixes the geometry of closure. When a system progresses through a cycle, it does not simply repeat, it returns with displacement. Orientation changes even as alignment is restored, and that change is phase. Cycles therefore produce continuity with difference, so while seasons recur, tides return, cells divide, history continues to accumulate. Without structured return repetition collapses into sameness or drifts into noise, but with it systems can oscillate while retaining structure.

Phase turns change into space and repetition into time. As phase advances, differences emerge, and those differences are what we call distance, while space is the bookkeeping of persistent phase differences. Time appears because phase carries orientation and order, allowing the system to distinguish before from after. Space may be viewed as frozen phase and time as flowing phase, with both arising from ordered repetition. Memory then emerges when phase relationships persist across interaction. When processes become correlated, their relative phase constrains future evolution. In this broad sense entanglement becomes any situation where phase relationships cannot vary independently, and once phase is stabilised the system carries history forward. Hence memory is seen as phase that survives resolution, and geometry is its expression as stable patterns of distance, angle, and symmetry.

As structure accumulates, entanglement increases both possibility and constraint. Early on degrees of freedom dominate, because new combinations are entropically cheap, correlations proliferate, so complexity grows rapidly. This is the ascending limb of the complexity axis, where however each additional layer of correlation is bounded by history. Constraint thereby accumulates until it begins to dominate, and the turning point occurs when freedom and constraint are balanced. This is the edge of chaos, where systems are most capable of adaptation and learning, whilst beyond it additional correlation reduces usable complexity. The system becomes over-constrained and brittle. Degrees of freedom remain, but they are locked into fixed relationships, and variation can no longer propagate, so thereby complexity recedes.

After peak complexity, entanglement becomes limiting rather than enabling. The system sheds correlations to regain manoeuvrability. Complexity is therefore sinusoidal, it rises, peaks, and recedes according to the balance between freedom and constraint. This applies across scales, from particles to ecosystems, brains, and cosmology, the same curve appears in lives and organisations. Growth and learning give way to increasing constraint through specialization and accumulated structure, followed by rigidity and decline. At the universal scale, the logic is always identical. Entanglement builds structure across space and time, and as correlations deepen degrees of freedom become increasingly constrained. Eventually constraint dominates, and the system resolves excess certainty through collapse. Therefore when constraint begin to dominate, relationships can no longer rearrange freely, and the system responds by curving its relational space. That curvature is what we observe as gravity. It is the geometric expression of retained inference under constraint. Where memory is light, space is flat, and where memory is dense, space curves. In this way objects follow geodesics because those paths minimise additional constraint.

A black hole represents the limiting case, where memory is compressed to the point that no further relational freedom is possible. Phase can no longer propagate outward, and geometry closes in on itself, so gravity becomes memory at saturation. It is universal and cumulative because memory aggregates and constraint compounds. Curvature is the shape of stored history, and phase is the minimal mechanism by which difference can persist. When phase relationships survive interaction, something is remembered, and that persistence becomes memory. Therefore once memory exists, constraint follows automatically, because retained relationships limit what can change next. Those limits give rise to geometry, which is the stable bookkeeping of relational constraints. From geometry comes structure, from structure comes complexity, and from complexity comes life, cognition, and cosmology.

Phase is also the mechanism by which uncertainty resolves. A heartbeat is phase made biological, where each beat is timing held in relation to the last, so that when phase coherence is lost, function degrades. The same applies to circadian rhythms. Systems entrain to shared cycles, and coherence emerges without instruction. Uncertainty presents many possible states, phase selects which relationships persist across cycles, because what aligns reinforces, and what does not fades. The system therefore becomes fundamentally biased towards persistence. In this sense, the universe refines its own trajectories under constraint. Geometry encodes stable relations, particles encode persistence, life encodes efficient inference, and gravity encodes saturation. Learning requires only that outcomes influence what follows, and phase makes that possible. When phase relationships persist, consequences are retained, and that retention constrains future evolution. The universe "learns" because phase allows the past to matter. The observer in wave collapse is phase. Before collapse, multiple phase relationships coexist, and collapse occurs when one configuration becomes stable enough to persist. That persistence is observation. Phase interactions accumulate until alternatives can no longer propagate. So the 'observer' is the moment phase stops being negotiable. Collapse is irreversible because phase ordering is irreversible once memory is encoded. The arrow of time appears at the same moment as observation, both arising from phase locking under constraint. What appears as nonlocal "action" is shared phase resolving coherently. Entangled systems already carry the same constraint, so collapse is a global update of a shared state. Space reflects the outcome because it is downstream of phase.

In a purely real space, information is limited to magnitude, such that histories are lost when values coincide. The complex domain adds phase as an independent degree of freedom. States can share magnitude yet remain distinct through phase, allowing more history to persist. The complex plane therefore increases memory capacity by preserving relational information across cycles. This is why wave mechanics, biology, and control systems converge on complex representations, because they allow systems to hold uncertainty, compare possibilities, and constrain outcomes before resolution. The complex domain is a higher-capacity memory substrate, enabling structured possibility to persist long enough for inference, adaptation, and emergence to occur.

Triads and Memory

Memory also shows up as structure within the standard model in the form of charges, spins, flavours, and gauge symmetries. They are the geometric expressions of what phase relationships are viable under selective pressure. Interactions conserve what must be remembered and forbid what would erase coherence. So the sequence holds that entangled inference creates correlated phase, and correlated phase persists as memory. Persistent phase demands geometric expression, and geometry stabilizes as particle properties and interactions. Structure then is the inevitable shadow cast by phase-retaining inference, and once that is understood, the standard model starts to appear as the memory lattice that reality uses to keep itself coherent over time. Moreover that encoding is in written in triads, whether in the standard model as color, or the codon and polyamines as structure, because triads are the smallest structure that transfer and carry forward memory.

While pairs can balance or oppose, they cannot provide direction or preserve sequence. With three elements however, it becomes possible to encode a before, during, and after, or reference, deviation, and correction. That is the minimum structure that allows history to persist through change. In the standard model, triads appear as SU(3) colour, as three related states allow confinement without collapse. The system cycles internally while preserving coherence, so information is not lost when interactions occur because the third element carries relational context.

In biology, the codon is a triad for the same reason. Two bases are insufficient to specify reliably, whilst three allow redundancy, error tolerance, and directionality. The genetic code remembers what worked while permitting variation. That is memory transfer across generations, much more than just chemical reaction. Polyamines show the same logic at a molecular scale. The triadic spacing stabilizes DNA and RNA by preserving phase relationships along the backbone, thereby preserving the geometry, and in that way allows information to remain readable. So triads matter because they support phase without stasis, they allow systems to change state while retaining relational identity, and that is exactly what memory requires. Across physics, biology, and chemistry, triads keep reappearing because reality needs a minimal structure that can survive interaction without erasing context, and in that context this repetition becomes a universal solution to the problem of retaining inference through time. Where color is the static phase which creates structure, flavor is flow which creates time, where ε is the ordering and the Planck constant is the boundary constraint.

In SU(3), color (as static phase) creates structure because it locks relational alignment, it does not progress or decay. It cycles internally while remaining confined, which means phase is preserved rather than spent. Preserved phase manifests as stable geometry. Hadrons exist because color phase cannot unwind, so structure becomes static memory. By contrast flavor (phase in motion) changes only through weak interactions that are directional. Once a flavor transition occurs, the system cannot be reversed without altering the history, and that irreversibility is what gives time its arrow. Flavor is ordered change and that ordering is what turns sequence into duration, where ε functions as the ordering principle. It is the infinitesimal that decides which way resolution proceeds when symmetry is otherwise perfect, so ε is the seed of ordering that allows inference to commitment, rather than perpetual oscillation.

The Planck constant acts as a boundary constraint, setting the minimum action required for a phase change. Below this scale distinctions collapse, so only above it can memory survive under interaction. Quantization therefore follows as the cost of retaining phase across resolution and taken together, the system is coherent. The ordering coordinate ε defines direction through resolution, while the Planck constant limits that resolution, enforcing discreteness so that each element performs a distinct role without overlap. Such separation is expected if the universe encodes memory through constrained inference rather than assembling reality from independent parts. Once stable structures emerge, entanglement generates complexity and constraint along the same axis. As long as degrees of freedom exceed constraint, complexity rises toward a maximum. Beyond that point, as previously discussed, constraint dominates and complexity recedes. In simple terms, a system needs at least three elements to preserve history through change, and with the triad a system can reference itself, so one element can change while the other two preserve context. That makes persistence possible and that persistence is memory. This is why triads recur wherever memory must survive interaction.

Emergence therefore is triadic in function and what determines emergence is the balance between three roles that must all be present at once:

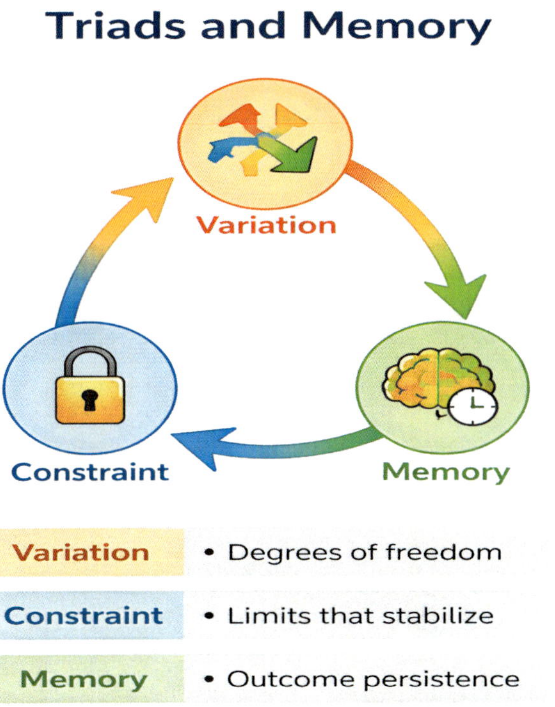

- **Variation**: degrees of freedom that allow exploration
- **Constraint**: limits that prevent dissolution
- **Memory**: persistence that carries outcomes forward

The triad allows one component to change while two stabilize the relationship, and this creates a ratchet. Change can occur without loss of identity and once that happens, emergence becomes possible because the system does not then reset each cycle. Along the complexity axis, emergence occurs precisely where these three are in balance. Too much variation and structure never stabilizes, whilst too much constraint and exploration dies. Too little memory and nothing compounds, so the peak of complexity appears where the triad is optimally tuned. This is why emergence clusters at the edge of chaos because past the peak, constraint overwhelms the triad. Degrees of freedom are still present, but they are locked and memory saturates. The system can no longer reorganize, so complexity retreats, emergence is gated and that gate is triadic. Wherever you see sustained novelty, adaptation, or life, you will find a triad quietly doing the work, something changing, something holding, and something remembering. That is what sets the position on the complexity axis, and why emergence appears sharply rather than gradually.

For common events size increases gradually and predictably, but beyond a threshold rare events grow much faster than expected, so small changes in rarity produce very large impacts. This image below illustrates a tipping point where linear behaviour breaks and extremes dominate, a hallmark of systems approaching their limits. These thresholds are predictable in principle, because emergence and collapse occur when a system crosses a balance point between three quantities that can be tracked, namely the available degrees of freedom, accumulated constraint from correlation and memory, and the efficiency with which memory is retained through phase. When freedom grows faster than constraint, complexity rises. When constraint catches up and overtakes, the system approaches a maximum. The threshold sits at the point where added correlation stops producing new usable structure and starts producing rigidity.

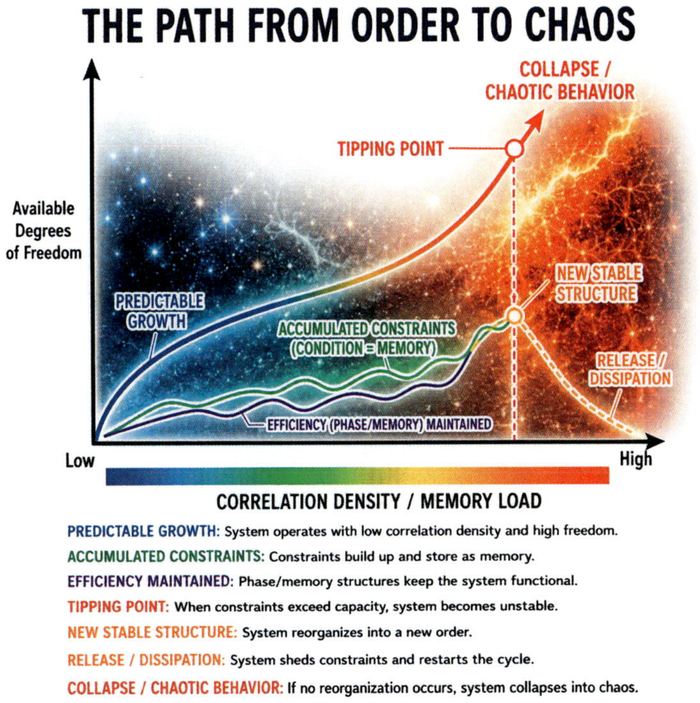

This behaves like a phase transition and we do not need to know every microstate. Once correlation density, coupling strength, or memory load crosses a critical band, the system's behaviour changes qualitatively, which is why the transition feels sudden even though the drivers were accumulating gradually. The exact location of the threshold depends on boundary conditions. Biology, organisations, brains, and cosmology each have different tolerances for constraint. Still the form of the threshold is universal, and it always appears as a sharp change in adaptability, learning rate, or response diversity. So although we cannot always say when collapse or emergence will occur to the day, we can say that it must occur once certain relational densities are exceeded and that is enough to make the process lawful rather than accidental. In short, thresholds are set by how much memory a system can carry before constraint overwhelms freedom. Once that capacity is approached, emergence peaks, and beyond it retreat is unavoidable.

This chapter has tried to establish a simple central result, that a universe that builds structure must be able to retain unresolved relationships across change. That capacity is memory, and memory cannot exist in a strictly Markovian description. Once history is allowed to persist, dynamics alone are no longer sufficient. The universe requires an internal degree of freedom that can carry orientation, ordering, and relational context forward without adding new forces or particles. That requirement already constrains the mathematics that can describe reality. Memory is stored as relative position within a cycle and what survives interaction is oriented with respect to what came before, and that orientation is phase. Phase is the minimal structure that allows difference to persist without freezing change, which allows repetition without erasure, correlation without collapse, and sequence without an external clock. If memory is physical, then phase must also be physical, because there is no cheaper way to preserve relational history.

Once phase is admitted, the real number line is no longer sufficient. A real value can describe how much, but not how it is aligned. To carry phase without duplicating variables or inventing auxiliary rules, the system must step off the real line into the complex plane. The complex field is the smallest extension of the reals that can store phase, support rotation, allow interference, and preserve history compactly. Any universe driven toward parsimony will converge on it. Chapter 2 begins at that point. Having shown that memory is unavoidable, it asks what mathematical structure can carry memory with the least possible cost? The answer is the complex number field. From there, interference, oscillation, chirality, spinors, twistors, holography, and colour appear. They become successive expressions of phase-retaining structure as recursion deepens and complexity scales. The complex world is the first place where a universe that remembers can exist at all.

Technical Appendix Chapter 1

1. Complexity

1) Scale with Entropic Potential

Let s be a scale (coarse-graining level). Let X_s be the effective configuration variables at that scale (field modes, molecular states, network states, etc.). Define an entropic potentiality $\Phi_s(x)$ and the induced distribution;

$$p_s(x) \;=\; \frac{e^{\Phi_s(x)}}{Z_s}, Z_s = \int e^{\Phi_s(x)}\, dx.$$

This lets us compute entropies and correlations as geometric properties of Φ_s.

2) Degrees-of-Freedom Density $d(s)$

A useful operational definition is effective dimension, via entropy relative to a reference resolution.
If X_s is discretised at resolution Δ(or binned):

$$d(s) \;\equiv\; \frac{H(X_s)}{\log{(1/\Delta)}} \text{(dimensionless)},$$

where $H(X_s) = -\sum_x p_s\,(x)\log\, p_s(x)$(or differential entropy in the continuous case).

Interpretation: how many independent "free" coordinates are actually being explored at scale s.

3) Constraint Density $c(s)$

Constraint is "non-factorisability" (dependence structure). 'Total correlation' (multi-information):

$$c(s) \equiv \sum_{i=1}^{n} H(X_{s,i}) - H(X_s).$$

If the variables are independent, $c(s) = 0$. As correlations tighten, $c(s)$ rises. This is the right object for "entanglement as constraint" and it counts the reduction of accessible microstates caused by binding.

4) Memory Density $m(s)$

Memory is 'persistence of constraint across updates'. Define two successive recursion steps (or update index) $r$and $r + 1$. Let the coarse state be $X_s^{(r)}$. Then a clean memory metric is mutual information across recursion:

$$m(s) \equiv I\left(X_s^{(r)}; X_s^{(r+1)}\right)$$

$$= H(X_s^{(r)}) - H\left(X_s^{(r)} \,\middle|\, X_s^{(r+1)}\right).$$

High $m(s)$ means the system carries forward structured information at that scale, not just noise. To make the "phase" aspect explicit, defining a phase-like coordinate θ_s(U(1) fibre coordinate);

$$m_\theta(s) \equiv I\left(\theta_s^{(r)}; \theta_s^{(r+1)}\right),$$

or (continuous phase) the circular correlation $\langle e^{i(\theta_{r+1}-\theta_r)}\rangle$as a coherence proxy.

5) Complexity Density $\boldsymbol{k(s)}$

A simple DE-consistent choice is 'usable freedom after constraint', weighted by persistence:

$$k(s) \ \equiv\ [d(s) - \lambda\, c(s)]_+ \ \cdot\ \sigma(m(s)),$$

where $[\cdot]_+ = \max(0,\cdot)$ and σ is any increasing saturating function (e.g. $\sigma(m) = 1 - e^{-m/m_0}$).
So complexity is high when we still have room to vary, correlations have not over-constrained the space, and what we build persists.

6) The threshold condition

Complexity rises to a maximum then retreats, which becomes a condition on the balance between DoF growth and constraint growth. The turning-point condition at the peak scale $s^{*}$is:

$$\frac{d}{ds}k(s)\ |_{s=s^{*}} = 0,$$

which, under mild assumptions on $\sigma(m)$, reduces to the intuitive rule:

$$\frac{d}{ds}d(s)\ \approx\ \lambda\,\frac{d}{ds}c(s)\text{at } s = s^{*},$$

with decline occurs when

$$\frac{d}{ds}d(s)\ <\ \lambda\,\frac{d}{ds}c(s).$$

So the peak occurs when new degrees of freedom arrive no faster than new constraints.

$$\text{Complexity}(s) \;\propto\; [\text{Freedom}(s) - \text{Constraint}(s)]_+ \times \text{Persistence}(s),$$

with freedom $= H(X_s)$, constraint $= \sum_i H(X_{s,i}) - H(X_s)$, persistence $= I(X_s^{(r)}; X_s^{(r+1)})$.

2. Electromagnetism

In the (DE) frame electromagnetism can be reinterpreted as the physical manifestation of an entropy gradient. An entropy gradient means spatial variation in accessible microstates:

$$\nabla S \neq 0$$

In thermodynamics, systems move toward higher entropy because configurations with more microstates are statistically favoured, but in information language, this is movement toward higher degeneracy of description. There are established examples where apparent "forces" emerge from entropy gradients:

- **Polymer elasticity** (entropic spring)
- **Osmotic pressure**
- **Verlinde's entropic gravity proposal**

In these cases:

$$F \sim T\nabla S$$

So the conceptual move is that the force is not fundamental, it is statistical. Electromagnetism arises from local U(1) phase invariance, formally;

- Gauge connection A_μ
- Field strength tensor $F_{\mu\nu}$

The electric field:

$$\mathbf{E} = -\nabla\phi - \partial_t \mathbf{A}$$

The magnetic field:

$$\mathbf{B} = \nabla \times \mathbf{A}$$

This is geometric curvature of a phase bundle, and no entropy term appears in the Lagrangian. But if memory and entropy resolution are primitive, then phase encodes memory persistence, charge represents topological phase asymmetry, and field lines represent constraint propagation. Then an entropy gradient may be seen as the spatial propagation of phase-memory imbalance. A charge then creates a constraint in phase space, and a local reduction in accessible configurations, with compensatory gradients in allowable states. The electric field then becomes:

$$\mathbf{E} \propto \nabla(\text{phase degeneracy})$$

In this sense electromagnetism is seen as the geometry of entropy redistribution in phase space. For the idea to hold, it must explain:

1. Inverse square law
2. Lorentz invariance
3. Photon as massless spin-1 excitation
4. Gauge symmetry

The inverse square law can emerge from 3D geometric spreading of constraint information, where Gauge symmetry is local phase redundancy. If phase is memory orientation in DE, then U(1) symmetry may be considered invariance under local memory relabelling, where the gauge field ensures consistent phase comparison between points. Electromagnetism in this frame is the dynamical response of phase memory structure to spatial entropy imbalance.

1) Minimal DE dictionary

We want a mapping that keeps orthodox physics intact, while reinterpreting the objects.

- **Memory phase:** $\theta(x)$
- **Local phase-relabelling freedom** (DE: "memory gauge"): $\theta(x) \to \theta(x) + \alpha(x)$
- **Connection** (tracks phase comparison across space, DE: "entropy bookkeeping field"): $A_\mu(x)$
- **Curvature** (phase mismatch around a loop, DE: "residual unresolved memory curl"):

$$F_{\mu\nu} = \partial_\mu A_\nu - \partial_\nu A_\mu$$

Core invariant combination:

$$D_\mu\theta \equiv \partial_\mu\theta - qA_\mu$$

This is the quantity that is meaningful physically: not $\partial_\mu\theta$or $A_\mu$separately.

2) Start from the smallest action that can work

Take a "memory medium" (a complex order parameter) but keep it stripped down. Let;

$$\psi(x) = \sqrt{\rho(x)}\, e^{i\theta(x)}$$

Use the standard gauge-invariant action (this is the smallest one that will automatically give Maxwell):

$$\mathcal{L} = -\frac{1}{4}F_{\mu\nu}F^{\mu\nu} + \frac{\rho}{2}\,(\partial_\mu\theta - qA_\mu)(\partial^\mu\theta - qA^\mu) - V(\rho)$$

DE reading:

- $V(\rho)$sets the "memory density" preferred state
- $(\partial\theta - qA)$ is the entropy-gradient-like drive in phase-memory space
- $A_\mu$exists because comparisons of memory between points must be consistent

If you want absolute minimalism, freeze $\rho = \rho_0$(stiff memory density), which reduces to:

$$\mathcal{L} = -\frac{1}{4}F_{\mu\nu}F^{\mu\nu} + \frac{\rho_0}{2}(\partial_\mu\theta - qA_\mu)^2$$

3) Derive Maxwell from the action

Vary with respect to A_μ we get :

$$\partial_\nu F^{\nu\mu} = j^\mu$$

where the current comes out as

$$j^\mu = q\rho\,(\partial^\mu\theta - qA^\mu)$$

That is already the Maxwell equation with a conserved source.

Now vary with respect to θ:

$$\partial_\mu(\rho(\partial^\mu\theta - qA^\mu)) = 0$$

which implies current conservation automatically:

$$\partial_\mu j^\mu = 0$$

So we can recover the core structure:

- Maxwell equations
- conserved charge/current
- gauge invariance

4) DE Entropy Gradient

The drive term is:

$$\Xi_\mu \equiv (\partial_\mu\theta - qA_\mu)$$

In DE terms, interpret θ as a local memory coordinate, and Ξ_μ as the residual mismatch between memory labelling and the connection that transports it. Then the "field" is not primary, it is the relaxation geometry required to reduce unresolved mismatch.

Electric field as DE-gradient

In temporal gauge language, the electric field is largely the gradient part:

$$\mathbf{E} = -\nabla\phi - \partial_t \mathbf{A}$$

In DE, that's the system reacting to spatial-temporal mismatch in θ transport. Charges are places where θ cannot be globally smoothed without a defect.

5) Inverse-square law

For static configurations (no time dependence) and choosing a gauge where θ is time-independent, Maxwell reduces to:

$$\nabla \cdot \mathbf{E} = \frac{\rho_{\text{charge}}}{\varepsilon_0}$$

Outside sources, $\nabla \cdot \mathbf{E} = 0$, plus symmetry gives:

$$E(r) \propto \frac{1}{r^2}$$

DE reading: constraint information spreads over spheres in 3D.

6) Magnetic field in DE is curl of constraint flow

Magnetism appears when the phase mismatch is not purely gradient-like, so you get circulation.

$$\mathbf{B} = \nabla \times \mathbf{A}$$

Effectively rotational memory transport or "twist" in the bookkeeping connection. So EM is the manifestation of informational/structural entropy gradients in phase-memory space, where the vacuum remains unitary and conservative because this "entropy" is not heat, it is degeneracy of compatible micro-configurations under constraints.

Emergence of Photons

Step A: Linearize around vacuum:

- set $A_\mu = 0 + a_\mu$
- small fluctuations in the gauge field

  Then the Maxwell kinetic term $-\frac{1}{4}F^2$ gives massless waves with two polarizations.

Step B: Pin charge to topology

Impose that θ cannot be globally single-valued around a defect:

$$\oint \nabla\theta \cdot d\ell = 2\pi n$$

That gives quantization conditions and ties "charge" to topological winding in the DE memory phase sector.

Chapter 2: The Complex World

If phase is physical then the real number line is no longer a sufficient expression, as the system must represent both magnitude and orientation. The minimal extension that allows this is the complex field, because it carries more information per degree of freedom, which expands the solution space while keeping the underlying rules compact. Where real number gives one degree of freedom, a complex number provides two. Magnitude and phase encode more structure, so that a field written in $\mathbb{C}$ can store amplitude and orientation without extra variables. This matches DE's pressure toward parsimony, which maximise information per symbol, whilst minimising rules. Complex structure allows reversible evolution, and unitary evolution in quantum mechanics needs a phase space that can rotate. The complex plane provides the simplest structure that supports this, whilst real-only formalisms end up recreating complex behaviour through paired variables anyway. Therefore a system that must retain both magnitude and orientation cannot remain on the real line.

Complex numbers stabilise oscillations, such that wave behaviour, interference, boundary conditions and conserved currents emerge cleanly when solutions can move through the imaginary axis. Without that axis, the same physics becomes convoluted or impossible. Complex structure doubles the microstate capacity without doubling the rule set, increasing entropy-resolution bandwidth without increasing the action cost. That makes the system more capable of finding low-entropy attractors. Complex numbers also make analytic continuation possible. This single feature binds quantum mechanics, thermodynamics, relativity and field theory. The *i-axis* permits movement between domains (Euclidean, Lorentzian, thermal, tunnelling) with no new axioms. Once again maximal reach for minimal extra structure.

Chirality and time-orientation require a sign structure, and the imaginary axis provides the cheapest way for a system to encode handedness and temporal orientation. Without the imaginary axis we would need larger groups or extra fields, and complex numbers are the lowest-cost carrier of predictive certainty, because they compress the most microstate information into the smallest formal object that still allows smooth evolution and stable attractor formation. Nature 'uses' complex numbers because they maximise informational reach for minimal mathematical cost, giving the universe more possible solutions with fewer rules.

Interference is only possible when states have a phase. Because real-only systems encode amplitude but must encode orientation indirectly, their superpositions cannot combine, cancel, or form structured patterns. Without interference, we lose selection, differentiation and emergence. Complexity needs recursion that can rotate and move forward and back in a line. Complex recursion moves around a plane, and that rotation produces spirals, bifurcations, attractors, and the full range of chaotic behaviour. Without that complex plane no stable chaotic geometry can emerge. Complexity also needs internal memory and phase stores history, as it encodes a previous state without extra variables. That memory capacity is the basis of constructive emergence. By removing the imaginary axis and we remove that memory channel.

Chirality, handedness, orientation and time direction all require a sign structure that real numbers alone cannot supply. Complex numbers supply the simplest symmetry-breaking unit. Complexity needs analytic continuation, moving between dynamical regimes, tunnelling, bifurcations and transitions all relying on the ability to shift through imaginary domains. Without that systems become trapped in local behaviour and cannot explore a richer landscape. Where real numbers describe things, complex numbers permit things to evolve, so that complexity appears the moment the universe steps off the real line and into the complex plane.

The Schrödinger equation drives the waveform through a landscape of amplitudes and phases. That evolution creates the interference patterns that seed structure and make complexity possible. The equation is a rule for how information moves, by letting the waveform shift, rotate and mix its components. This motion produces patterns rather than noise, interference that creates structure. Where phases align, we get reinforcement, where they oppose, we get cancellation. This is the first step toward ordered form, whilst recursion through complex phase produces emergent detail. Each small step of evolution adds new structure that depends on the previous state. This recursive layering is the core of complexity. As the waveform holds both possibility and history simultaneously, it's the phase record guides future evolution, and it is that stored information that lets structure persist and grow.

Informational Superiority of the Complex Number Field in Cartesian Representation

Mathematical formalisms serve as the scaffolding that supports physical description, so the choice of numerical domain determines how much structure a theory can express. Complex numbers occupy a privileged position because they provide a richer informational basis than real numbers while retaining a compact rule set. When mapped into Cartesian form as ordered pairs, they reveal why the complex field supports deeper dynamical structure, more detailed geometry, and more capacity for emergent behaviour.

1. Complex Numbers as a Two-Dimensional Information Carrier

A real number occupies a single dimension, as it specifies magnitude without orientation. A complex number recorded as the pair (x, y) occupies a plane. The field inherits linearity and algebraic operations from the reals yet expands the informational bandwidth by introducing a phase dimension. This gives each point an amplitude plus a rotational degree of freedom. The second degree of freedom does not require additional axioms, it is supplied by the imaginary unit operating as a ninety-degree rotation. A Cartesian complex plane therefore stores twice the information per coordinate while preserving simplicity of transformation. This density of representation means that complex fields supply a minimal extension of the reals that yields a non-trivial expansion of expressive power.

2. Geometry Enhanced by Phase

Geometric relations expressed on the real line are limited to ordering and distance. Expressed on the complex plane they extend to rotation, orientation, curvature, holonomy, and conformal structure. A point on the complex plane carries direction relative to the origin as well as magnitude. This orientation allows richer descriptions of flow, oscillation, and recursion. Complex coordinates therefore represent vectors, transformations, and flows more naturally. Real coordinates require additional structure to express the same behaviour. The informational advantage arises because phase stores orientation in the minimal possible form.

3. Natural Encoding of Oscillatory and Recursive Processes

Any oscillatory phenomenon requires a pair of coupled parameters. Real formalisms must introduce these parameters separately, whereas the complex field unifies them automatically. A wave written as $\mathrm{A}e^{i\theta}$ stores amplitude and phase in one coordinate. The Cartesian representation $(x, y) = (A \cos\theta, A \sin\theta)$ reveals two conjugate components evolving coherently under the same rule. The informational efficiency is high because the rotation operator *i* acts as a uniform generator of oscillation. Recursive processes such as the Schrödinger evolution or classical wave propagation become compact when expressed in complex form. Each step of evolution updates both coordinates in a coherent manner, which strengthens the capacity for pattern formation. No separate bookkeeping is required to preserve the integrity of the paired components.

4. Complex Differentiability as an Information Filter

Complex differentiability imposes stronger constraints than real differentiability. The Cauchy-Riemann conditions enforce a strict relationship between the Cartesian components. A function that satisfies them is determined completely by values on any open set, and this rigid structure allows powerful theorems such as analytic continuation. The field therefore acts as an information filter where redundant descriptions collapse, while lawful patterns survive. In informational terms the complex plane reduces noise by enforcing coherence between $x \; and \; y$. Real functions lack this constraint and require external structure to achieve similar filtering.

5. Conformal Symmetry and Preservation of Local Structure

Transformations on the complex plane preserve angles, local ratios, and shape. This feature allows the complex field to encode the structure of flows without distortion. Real functions that attempt to model equivalent transformations lose information or require more elaborate representation. When expressed in Cartesian form, this symmetry is clear. The two components rotate together under the mapping $z \rightarrow \mathrm{e}^{i\theta} z$, where orientation is preserved without shear. This behaviour explains why complex analysis is central to field theory, potential flows, probability amplitudes, and dynamical systems that require precision in local geometry.

6. Complex Multiplication as a Compact Geometric Operation

Multiplication by a complex number performs a simultaneous scaling and rotation. Performing the same operation in real coordinates requires a two-by-two matrix. The complex field compresses that matrix into a single number. The mapping from algebra to geometry becomes direct. In informational terms this reduces overhead. A single instruction carries a transformation that would otherwise require four parameters. This compactness is a reason why quantum mechanics and signal processing use complex fields as their natural algebraic base.

7. Expansion of the Solution Space

Many physical equations written over the reals lack complete solutions. The extension to complex coordinates closes the solution set. This closure property increases informational reach. A complex solution supplies amplitude plus phase. Even when the physical observable is real, the pathway of evolution often requires the richer internal coordinate. As a Cartesian pair, the complex coordinate reveals how the dynamical system explores a two-dimensional internal landscape to produce stable real outcomes.

In summary a complex number treated as a Cartesian coordinate stores more information per symbol, supports richer geometry, enables oscillatory and recursive processes, and preserves structural coherence through strong differentiability constraints. These features arise without expansion of the foundational rule set. This economy of structure paired with increased informational bandwidth explains why complex fields form the backbone of modern physics. They provide a minimal mathematical extension that yields maximal descriptive capacity. Real systems that attempt to capture the same detail require additional variables or auxiliary rules. The complex plane delivers those features natively. In an evolutionary universe, the adoption of complex structure becomes inevitable. The complex field enables greater informational reach with fewer rules, so any system driven by parsimony will settle on it as the natural basis for generating structure and complexity, and evolution favours efficient information carriers.

Complex dynamics support coherent memory, because phase carries history in a compact form. This allows systems to accumulate information across steps of evolution. Real systems need larger state descriptions to retain a comparable record. A universe that evolves toward rule-efficiency will incorporate complex phase as the default medium of memory. Complexity requires interference, interference patterns seed structure, structure requires phase, and phase requires the complex field. This is the spiral of emergence, and without complex numbers the pathway from simple rules to emergent structure becomes obstructed, so evolutionary selection removes that obstruction by adopting the complex plane.

Complex representation reduces the cost of transformation. Scaling and rotation reduce to a single operation, so patterns can evolve and fold through a richer landscape with minimal instruction length. This drives a lower action cost in any evolutionary process, which aligns with parsimony as a selection principle. This is why topology and deductive entropy demand the complex field, because without the complex number field matter would not emerge. The complex field introduces a second internal coordinate that allows waves to combine, cancel, rotate, and store history. These behaviours produce standing patterns, which become the precursors of particles. Phase coherence defines identity and separation, interference defines localisation, and recursion on the complex plane defines stability. Without that internal phase, waves would not lock into persistent forms, they would disperse without structure. The complex field provides the minimal machinery that lets patterns harden into matter. Mathematics often discovers ideas long before their significance becomes clear, and few examples illustrate this better than spinors and twistors. Both objects now sit at the heart of modern physics, yet neither was introduced because anyone was looking for a new building block of nature. Instead, they were invented to solve much narrower problems, and only later did physicists realise they were touching something deeper.

Spinors

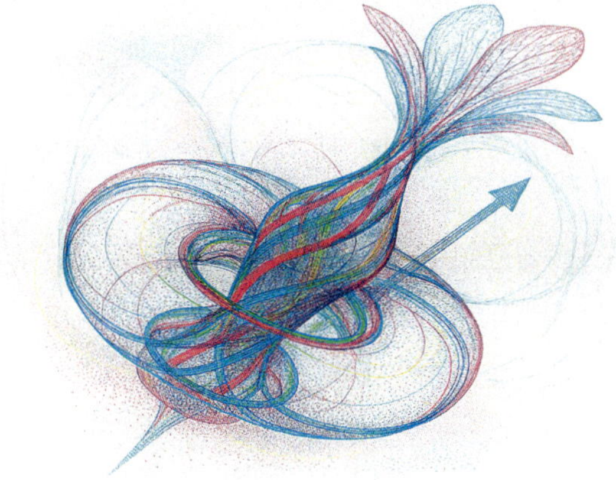

FOUNDATIONS OF COMPLEXITY
Information Recursion Structure (Deductive Entropy)

This object represents a single spinor. It is not a vector and it does not live in spacetime. Instead, it encodes orientation,

phase, and handedness in a form that only becomes geometric when paired with its conjugate partner.
The looping and self-intersecting structure reflects the defining property of spinors, where a full rotation does not return the object to itself. Closure occurs only after two complete turns, expressing the double-cover nature of rotation. The evident twist gives the object chirality, a built-in left–right distinction that cannot be removed without changing its identity. The flowing layers and colour variation represent phase rather than position, or the information that determines how this object combines with its partner to produce vectors, mass, and spacetime structure. On its own, the spinor is incomplete, as it carries potential rather than geometry. Only when two such objects are coupled does familiar spacetime emerge.

A spinor looks like an ordinary vector at first glance, but it is a pair of complex numbers arranged in a column. Spinors behave in a way that surprised every physicist who has encountered them as it only returns to itself after two full turns. This strange behaviour, discovered in the 1800s by William Rowan Hamilton and others working on the algebra of rotations, held no obvious physical meaning at the time. Spinors seemed like mathematical curiosities with no connection to the real world. That changed in the 20th century, when Paul Dirac wrote down his famous equation for the electron, he found that the natural mathematical object describing the electron's quantum state was indeed a spinor. Suddenly the "curious" object from long-forgotten algebra turned out to be the correct description of matter. This raised a deep puzzle because why would nature choose such a strange mathematical object to describe its most basic particles? To this day, there is no consensus on why the universe uses spinors, yet Physics uses them constantly.

A spinor becomes real by pairing with its conjugate within the same field. For the electron, this pairing is already present as its left- and right-chiral components. Geometry then arises from the internal coupling of phase and its conjugate structure. In this sense, the electron already contains the minimal dyad required for spacetime to appear.

Twistors

Twistors were introduced by Roger Penrose in the 1960s. It is a more elaborate structure that combines a spinor carrying orientation and phase, plus a geometric component representing position in space. Where spinors capture "how a particle points", twistors capture "how a particle moves", especially along paths of light. Twistors describe objects that travel at the speed of light (such as photons) far more naturally than ordinary spacetime coordinates. In twistor space, the paths of light become simple geometric lines rather than complicated curves.

FOUNDATIONS OF COMPLEXITY
Information Recursion Structure
(Deductive Entropy)

This object represents a twistor, which is best thought of as a strand of reality that links orientation, motion, and information before spacetime fully exists. The central braided column shows a tightly wound structure that carries direction and phase together, and as such it can be regarded as a rule for how space and motion can later emerge. The twisting indicates that rotation and direction are inseparable, and that a full return to the same state requires more than a single turn. The circular loops surrounding the core represent families of possible light-like paths. These loops are the set of directions along which influence or information could propagate if spacetime were resolved. Their symmetry shows that these possibilities exist before any particular location is chosen, and the upward arrow indicates emergent direction rather than movement. It marks how a consistent sense of time and causality appears once many such structures align and constrain one another. Time here is not assumed in advance, it is a consequence of repeated, coherent twisting. The colour variation represents phase and internal structure, which determine how this twistor can combine with others. When many twistors agree, spacetime points appear as intersections rather than as primitives. In simple terms, this image depicts a unit of pre-spacetime structure as a generator of geometry. But the puzzle that remains is that why do they exist at all? Why should the universe have a hidden twistor geometry? For more than a century, spinors and twistors have had three striking features:

1. **They work**
   They describe quantum particles, light, and fields exquisitely well.

2. **They appear everywhere**
   In quantum mechanics, relativity, particle physics, and even condensed matter.
3. **No one knows why**
   They do not arise naturally from spacetime geometry. They appear as if "bolted on," mathematically essential but conceptually mysterious.

What is it about the underlying structure of the universe that seemingly makes spinors and twistors inevitable? Any structure that must encode orientation, phase, and transformation rules requires a representation that is not reducible to vectors. In DE space is considered to be emergent from three geometric axes. Time is not regarded as fundamental, but is a projection of the recursion structure. Topology carries the real generative rules of recursion depth, complexity scaling, and phase-memory. This topological subsystem has properties that require a complex phase, orientation sensitivity, the ability to represent chirality, and the ability to encode transitions that flip orientation. These features are precisely what spinors naturally represent. In DE spinors are emergent, they appear when the topological structure is embedded into the emergent 3D geometric space. Once spinors exist, twistors arise as the natural way to combine the spinor (orientation, phase, chirality) and the geometric embedding (position, propagation). In DE, a twistor is the geometric expression of a topological information structure once a spatial manifold has emerged. This perspective clarifies something that puzzled Penrose for decades because twistors look fundamental, but only because they express the deeper topological information once geometry exists. Spinors represent topological orientation in emergent space and twistors represent the combination of orientation + propagation, but both are inevitable once the 3+3 DE structure is present.

In this framing spinors and twistors are the natural algebraic and geometric expression of this deeper structure. This clarifies why particles come in left- and right-handed forms, why massless fields are described by spinors, why photons are represented spinorially, why matter–antimatter asymmetry persists, and why twistor geometry encodes physical law so efficiently, because spinors and twistors are the visible part of the topological recursion structure that generates space, time, and physical law. The DE world is seen as a six-axis structure, where three axes define the geometric sector, from which space emerges. Three axes define the topological sector, which governs recursion, complexity, and phase. These topological axes determine how states transform, how structure accumulates, and how information is retained, and they shape the rules by which space becomes possible. The topological sector contains three essential features:

1. Recursion depth, the ordering of updates, perceived as time in spacetime.
2. Complexity scaling, how structure deepens.
3. Phase-memory, how orientation and identity are retained.

Within this topological subsystem, chirality (handedness or directional bias) arises naturally as a feature of the recursion rule itself. When the geometric axes stabilise into an emergent three-dimensional space, the topological structure needs an algebraic object capable of representing its orientation, phase, and transformation rules. The simplest such object is a two-component complex spinor. In this view, a spinor is the minimal algebraic representation of the topological subsystem when it is embedded in emergent space. This single insight dissolves many old puzzles, as spinors appear because the universe's underlying structure carries a complex phase, an orientation rule, and a recursion framework that cannot be captured by ordinary vectors. Spinors are the smallest object that can encode all those features, and once spinors exist, twistors emerge by pairing this topological representation with the geometric embedding of space. A twistor simply combines the information carried by a spinor, and the positional degrees of freedom of the emergent geometric manifold. This pairing produces the familiar twistor structure used to describe light, null directions, and massless particles. Therefore twistor geometry arises once both the geometric and topological sectors are present.

This perspective makes the role of spinors and twistors clear. Spinors encode the phase and orientation properties of the topological subsystem, whilst twistors encode how these properties project into the geometry of space. Both are necessary once the six-axis structure is embedded into a spatial manifold, neither needs to be assumed at the beginning, neither requires a pre-existing spacetime, and both reflect the deeper recursion structure that gives rise to space, time, and physical law. This offers a natural resolution to one of the longstanding mysteries in mathematical physics as to why the universe seems to use spinors and twistors so deeply. They are the visible shadows of the topological architecture that generates everything else.

In 4 real spacetime dimensions:

$$\mathrm{Spin}(3,1) \rightarrow SL(2,\mathbb{C})$$

This is the unique situation where A 4-vector can be written as

$$V^a = \sigma^a{}_{AA'}\, \xi^A \bar{\xi}^{A'}$$

And the spinor representation splits into two distinct 2-component sets:

left-chiral spinor: ξ^A
right-chiral spinor: $\eta_{A'}$

These are Weyl spinors and this splitting happens only in four dimensions (technically 3+1 or 4 Euclidean, once complexified). This is the key point, because in 4D, spinors factor into two inequivalent chiral representations, whilst in higher or lower dimensions, they do not.

- 4 dimensions is the only setting where chirality naturally appears.
- The algebra of rotations and Lorentz transformations picks out exactly two kinds of spinorial behaviour.
- This mirrors the handedness of the world (left/right helicity, matter/antimatter distinctions, etc.).

DE's 6-axis structure ontology fits this perfectly;

- **3 geometric axes** → emergent 3D space
- **3 topological axes** → recursion, complexity scaling, phase-memory

This 3 spatial + 3 pre-geometric structural degrees of freedom does not live in spacetime. This is consistent with the algebraic fact that spinor splitting happens only after selecting a 3+1 projection. DE's geometry is 3-dimensional because there are 3 geometric axes → Euclidean spatial extension. The DE geometric axes produce a 3-dimensional spatial manifold, isotropic and Euclidean at the basic level, without time and without Lorentzian signature. This is exactly what Penrose argues, namely that space is geometric, that time is not and the DE geometry agrees with that structure, where time is seen as a projection from recursion. Where traditional physics assumes 4D spacetime, DE says space is geometric and time is topological. This immediately preserves the 3D geometry required for the spinor splitting phenomenon while avoiding the conceptual problems of Minkowski time. Chirality also arises from the topological axes (from phase-memory + recursion inversion), not from geometric spacetime. Spinors emerge only when the 3 geometric axes + 3 topological axes are projected into 3+1 dimensions. This is the precise structural mechanism:

Step 1. Topological subsystem
(recursion, scaling, phase-memory) carries orientation and chirality.
Step 2. Geometric subsystem
(3 spatial axes) provides an embedding environment.
Step 3. Projection
maps topological information into geometric space.
Step 4. Minimal algebraic representation
of this projection is a 2-spinor.
Step 5. Algebra now guarantees
two inequivalent spinor classes, Left / Right (Weyl).

This is the unique embedding in which this structure appears naturally, and in turn this gives a deeper origin for the Weyl spinor splitting. So chirality arises from non-invertible complex phase accumulated across recursion cycles in a curved uncertainty field. Topology provides the stage, whilst recursion provides the asymmetry, and when entropy selects the branch, chirality is the consequence. When this structure is embedded into 3-space with an ordering projection the spinor representation emerges. This explains why the twistor pair exists and why 4 dimensions are special. The three geometric axes give rise to a three-dimensional spatial manifold, while the three topological axes encode recursion, complexity scaling, and phase-memory.

Chirality is a property of the topological subsystem and emerges as a direct consequence of the 3+1 projection. In this way DE proposes an account of why the universe may select the unique dimensionality in which spinors fall into left- and right-handed groups, and why twistor geometry arises as a natural representation of the same structure. Physical processes need an ordering, a way to tell one update from the next. This ordering comes from the topological recursion axis, which the mind interprets as time. Spinors, which are the minimal representation of orientation and phase, naturally fall into two distinct families only in a four-dimensional world. This is why Peter Voit says, "Four dimensions are special because the spinors break up into two groups," and no other dimensionality produces this behaviour.

Spinors and twistors are therefore the minimal structures required to encode orientation, phase, and transformation once memory is admitted as physical. They represent constrained possibility, the set of states that can persist without erasing what has already been established. In this sense, they are expressions of the underlying logic of the universe. A universe that retains its own history must organise that history, and the simplest way to do so is through structures that carry phase, enforce constraint, and allow evolution without loss of identity. Spinors and twistors are precisely those structures. They are the natural language of a universe that remembers.

1. Spinors as square roots of vectors

In relativistic geometry a spacetime vector is built from a pair of spinors. In 4D Lorentzian spacetime this takes the form;

$$v^{AA'} = \psi^{A}\,\bar{\psi}^{A'}.$$

A single spinor lives in a deeper structure space. Only when you take a spinor together with its conjugate does a vector appear. In that strict sense, a spinor really is a square root of a vector, not metaphorically but algebraically.

2. Chirality as Fundamental

Each solution of the Dirac equation decomposes into two Weyl spinors of opposite chirality. They transform under different irreducible representations of the Lorentz group. One is left-handed, the other right-handed. A single chiral spinor cannot generate a timelike spacetime vector on its own, only the dyad can do this. This already tells us something important, that spacetime vectors are composite objects, and that chirality is more primitive than spacetime.

3. The Dyad Universe Interpretation

From that perspective each chiral solution can be viewed as a partner in a dyadic structure. Spacetime then appears as the bilinear projection that stitches the two chiral sectors together. In this framing, a "dyad universe" is a single emergent spacetime arising from the mutual constraint between two chiral solution spaces that cannot individually represent spacetime reality.

This idea already has deep roots. Twistor theory (as developed by Roger Penrose) treats spinors as primary and spacetime as derived. The fact that massless particles live naturally in spinor space while massive particles require chiral coupling fits this picture, where mass becomes the bookkeeping term for how tightly the dyad is locked together. One subtle but important point is that the two chiral partners are mutually incomplete descriptions. Each carries phase, orientation, and constraint information that only becomes causal spacetime when paired. Separation destroys locality, time, and mass simultaneously. Seen this way chirality is a fundamental dyadic structure of reality, where vectors, time, and causality are secondary consequences of resolving that dyad.

The Holographic Principle

The holographic principle says everything happening inside a region of space can be fully described by information stored on the boundary of that region. It is a literal claim that says a 3D volume behaves as if it is encoded on a 2D surface, and 4D spacetime behaves as if it is encoded on a 3D surface. Black holes have entropy proportional to area (not volume) which tells us that information is fundamental, where dimensionality and spacetime is the encoding.

In DE the geometric axes create space, whilst the topological axes create information structure. The world is generated by recursion and projection, and time is the ordering of recursion. Spinors, twistors, and even spacetime geometry are representations of deeper topological information. If the holographic principle tells us information lives on a boundary and physical reality is a projection of deeper structure, then the boundary must be able to encode the interior, and it must be stable, finite, and recursively updateable. Therefore the topological axes hold all information, whilst the geometric axes hold only spatial extension.

Collapse always happens on a 2-dimensional surface because collapse is a boundary phenomenon. A boundary in a three-dimensional space is always two-dimensional. The topological subsystem holds the information and the rules for change, the geometric subsystem expresses that information as space. When recursion saturates, the topological rules project onto a 2D collapse layer. This is why horizons, wavefronts, and phase boundaries are always two-dimensional, and why the universe looks holographic from the outside while remaining three-dimensional within.

Holography tells us that bulk spacetime physics is encoded on a lower-dimensional boundary, so that information is encoded as spinors. Where a vector needs volume, a spinor only needs orientation and phase, and that is precisely the data holography preserves.

Dyads Explain Holography

If a vector is a bilinear of two chiral spinors, then bulk spacetime is already a projection. In other words, the boundary theory keeps the dyad unresolved, and the bulk appears only when the left and right chiral sectors are paired and locked. This reframes holography as deferred collapse where phase, chirality and conformal structure are retained, and metric volume, mass and proper time only appear when the dyad is completed as a bulk operation. So effectively the bulk is generated from the boundary surface.

The Holographic Principle already explains that geometry emerges from entanglement. Spinor dyads tell us that entanglement binds the two chiral halves into something vector-like, where strong entanglement yields timelike structure, and weak entanglement yields null structure. No entanglement yields no spacetime at all, and this makes spacetime a secondary bookkeeping device for correlations.

In fact Twistor space is inherently holographic. It encodes null rays, phases, and chirality without committing to bulk points. Hence the twistor is effectively the unresolved spinor dyad living on the boundary of spacetime. Spacetime points only appear when enough twistors converge. Hence holography works because reality is chiral at its roots, spacetime vectors are late-stage composites, and the boundary keeps the primitive objects. The bulk is what we get when the dyad is forced to close. Once viewed in this way holography becomes inevitable.

From Spinor Phase to Colour Phase

Spinors can be understood as the minimal way a system keeps track of orientation when a simple vector is no longer a sufficient expression. Unlike vectors, a rotation of 2π does not return a spinor to its original state. The system retains a record of having completed one full loop, and only returns to closure after 4π. This record is carried in the complex phase, and what appears as a mathematical peculiarity is in fact a form of memory. Chirality is the first indication that this phase is not neutral bookkeeping. Left- and right-handed spinors inhabit distinct complex structures, and the same phase that produces double-valued rotation also encodes a fundamental asymmetry.

Spin, chirality and gravitational redshift are all different faces of the same object, namely how phase accumulates along different paths. At this point it should not be surprising that the strong force is also a story about phase. Colour is a three-valued phase label, where the same complex numbers that gave us spinors and chirality now pick up a new role. They tell us why there are three colours, why the gauge group that describes them is SU(3), and why the characteristic angle is $2\pi/3$. Once we treat phase as a record of how uncertainty has been resolved along a path, the choice of gauge group becomes the minimal way to keep the recursion consistent.

Superposition

In textbook QM, a state is written as

$$|\psi\rangle = \alpha\,|0\rangle + \beta\,|1\rangle$$

Where geometrically $|0\rangle$and $|1\rangle$ define a two-dimensional complex. Superposition therefore is a structural capacity of any system that has not yet exceeded its constraint threshold. It is the condition in which multiple pathways remain viable because the system retains sufficient freedom to explore them without committing, where the system explores a two-mode manifold, amplitudes rotate continuously within that manifold, so phase relationships are defined. Mathematically superpositions are generated by SU(2) rotations, where ordering (without decoherence) is an

SU(2) action. So superposition become the generic description of behavior for SU(2)-governed systems.

Our dyad has the defining properties of oscillatory dynamics, reversibility, correlations without retained memory, and no irreversible commitment. So a dyad effectively appears as superposition, because time is emergent and sequences are order in phase. Inside SU(2) all evolution is unitary, because while amplitudes rotate and entanglement is reversible, there is no preferred direction in state space. Therefore true collapse cannot occur within the SU(2) regimen and something must break SU(2) symmetry for collapse to happen. That "something" is triadic lock-in through entanglement, leading to irreversible constraint, with the loss of degrees of freedom, along with the writing of memory into the system

A two-mode system maps naturally onto SU(2), where evolution is reversible and no memory is written. As long as the system remains within this regime, it can explore possible states without commitment. Collapse becomes inevitable only when additional constraints accumulate, forcing the system beyond this two-mode structure. At that point, reversible evolution fails, degrees of freedom are reduced, and a single outcome is selected. Superposition is therefore the natural condition of a system that has not yet reached its constraint limit, and Hilbert space can be understood as a representational scaffold for this structure rather than a fundamental substrate. When entanglement is low, subsystems are acting independently, perturbations remain local, and outcomes are not yet locked. That is operationally indistinguishable from "superposition" in the classical sense, it is a structural condition of non-commitment.

Thus the primitive object becomes a system oscillating between two feasible modes without irreversible commitment. That is our dyad and it applies equally to a qubit, a biochemical pathway, a neural decision, or a patient hovering near collapse. The mathematics may differ, but the structure is always the same. Collapse occurs when the system cannot remain dyadic, where buffering fails, and a third constraint forces commitment with removal of degrees of freedom. Collapse then is simply the loss of potentiality. In this frame collapse is structural exhaustion, where the entropic cost of remaining undecided becomes too great.

Every dyad is equivalent to a harmonic oscillator at the structural level, because the minimal reversible two-mode system is always dynamically equipoised, conservating of potential without the pruning of future possible pathways. In a dyad the system can return to its prior state because no history is written, trajectories are closed, the defining feature of harmonic oscillators. Harmonic oscillators cannot collapse as long as a system remains dyadic, because collapse requires anharmonicity. In this frame collapse occurs when buffering fails as constraints accumulate, so that the potential is no longer quadratic, introducing mode coupling, energy leakage, and irreversible deformation of phase space. Triad lock-in results when reserve exhaustion forces pathway pruning. Therefore what is traditionally called quantum superposition corresponds to persistence within this harmonic regime, and collapse occurs when accumulated constraints force the system into an anharmonic state, breaking reversibility and eliminating closed trajectories.

Collapse and Anti-Harmonic Growth in the Effective Hamiltonian

In the buffered dyadic regime, the effective dynamics reduce to two conjugate modes that exchange structure without net loss. We describe this with an effective harmonic Hamiltonian written in terms of an ordering parameter ε (the recursion index), not a pre-existing time coordinate:

$$H_0 = \frac{1}{2}(\frac{p(\varepsilon)^2}{m} + \omega^2 x(\varepsilon)^2)$$

The corresponding trajectories are oscillatory in ε:

$$x(\varepsilon) = A\cos(\omega\varepsilon + \phi)$$

The system cycles through configurations and returns to its starting point. No configuration is privileged, no record is written, ordering exists, yet history is not recorded. This regime corresponds to compensated biological systems as well as quantum superposition, where oscillation preserves optionality. The system can revisit prior states because nothing irreversible has occurred. As long as the Hamiltonian remains harmonic, trajectories in phase space are closed, so that perturbations are resolved by oscillation rather than commitment. The system neither accumulates memory nor selects outcomes. In this regime, ε is cyclic rather than consumptive. It orders updates but does not create an arrow of time. Geometry may later reinterpret this ordering as time, but at this stage no directionality exists, and time has not collapsed.

Constraint Accumulation as Anharmonicity

As buffering capacity is depleted, constraints accumulate. In Hamiltonian terms this appears as higher-order contributions to the effective potential:

$$H = \frac{1}{2}(p^2 + \omega^2 x^2) + \lambda_3 x^3 + \lambda_4 x^4 + \cdots$$

These terms represent irreversible commitments, coupling to additional constraints, loss of symmetry, and depletion of reserve. They do not merely perturb the harmonic motion, they destroy it so that harmonicity is broken, and once anharmonic terms dominate, trajectories no longer close. Phase space develops separatrices and sinks, oscillation no longer resolves perturbations, so that recovery becomes incomplete. Buffering now costs structure rather than preserving it, and in DE terms, feasible future states are pruned and the reserve is irreversibly consumed.

When the system transitions from harmonic to anti-harmonic dynamics, ordering becomes directional. Phase space volume contracts and trajectories flow toward attractors. Prior states cannot be revisited, and at this point ε becomes consumptive. Time is no longer a neutral ordering of updates, it becomes the irreversible consumption of degrees of freedom.

The triad as the minimal generator of collapse

The smallest structure capable of producing irreversible anti-harmonic behaviour is a triad. Two modes can oscillate, but three constraints cannot all be satisfied simultaneously. One degree of freedom must be sacrificed, so oscillation fails. The triad is therefore the minimal generator of time collapse and history formation. In the harmonic regime, reduced density matrices remain approximately factorisable, entanglement is transient, and Von Neumann entropy remains low. In the anti-harmonic regime, coupling dominates, factorisation fails, and entanglement becomes persistent, so that Von Neumann entropy rises sharply, converging at the loss of oscillatory freedom.

Collapse then is not an instantaneous jump, an external measurement, or a violation of unitary evolution, it is the loss of periodicity. Once closed trajectories cease to exist, the system must commit, because no alternative remains. In compensated systems, effective dynamics are dominated by harmonic terms, generating reversible oscillatory behaviour in which ordering is cyclic and no history is written. As buffering capacity is exhausted, higher-order constraint terms accumulate, breaking harmonicity and eliminating closed trajectories in phase space. This transition renders ordering directional and consumptive, producing irreversible lock-in to attractor states. Oscillation persists while it is affordable, when it is not, commitment is forced, and history is written when the price of possibility exceeds the price of certainty.

Chapter 3: The Standard Model

The Standard Model can be read in two very different ways. In the usual account, it begins with symmetry groups and ends with fields and particles. In DE I take the opposite route, beginning from the informational side and ask what structures are required if fermions are to obey the Pauli principle, form bound states, and close consistently under recursion. From that perspective, the Standard Model emerges as a stable attractor under entropic filtering.

Quantum Field Theory (QFT) describes the universe in terms of fields that extend through space, with particles appearing as local excitations of those fields. Interactions are understood as exchanges of these excitations, and in practice the theory is evaluated by calculating scattering amplitudes between well-defined initial and final states. Predictions are extraordinarily accurate when disturbances are small and local. Interactions can be treated as perturbations around a fixed background, and the field behaves as if it were transmitting forces between particles. The difficulty arises when the disturbance is no longer small. At that point, the background itself is reshaped. Perturbative methods fail, and behaviour must be recovered indirectly. This is a structural signal that suggests the field is storing deformation. The response of the system then reflects accumulated structure, as well as the instantaneous exchange between particles.

An elastic material provides a useful analogy. Near equilibrium, small displacements produce linear responses that are easy to predict. As strain increases, the response becomes nonlinear, history-dependent, eventually reorganising the material itself. The governing equations have not changed, but the interpretation has. The breakdown of perturbative field theory is perhaps a signal therefore, in that we are treating a memory-bearing substrate as if it were memoryless. When perturbation fails this memory becomes visible and in this light, fields can be better understood as maps of stored constraint, whose dynamics reflect how that constraint is redistributed under disturbance.

QFT struggles most in strongly coupled, history-dependent regimes. That is exactly where memory and accumulated constraint would be expected to dominate.. In those regimes, the idea that a field is a smooth bookkeeping device for instantaneous forces starts to look like the incomplete abstraction. Renormalisation then reflects the absorption of accumulated strain into effective parameters. Entanglement and vacuum correlations appear naturally if the substrate already carries structure. Prediction improves when the field is treated not as a force carrier, but as a medium that yields under constraint.

This shift in perspective has a direct consequence. If the field stores constraint, then the internal labels of the Standard Model must reflect the minimal structures required to preserve that constraint under recursion. The appearance of three symmetry groups is then better seen as the result of three distinct uncertainty manifolds that cannot be collapsed further without loss of predictive capacity. Renormalisation makes sense if short-scale disturbances permanently deform the medium and must be absorbed into effective parameters. If fields were forces, divergences would be pathological, but if fields are elastic substrates, renormalisation simply becomes coarse-graining stored stress. Entanglement and vacuum correlations look odd if influences must propagate force-like through spacetime, but they look natural if the field already contains pre-strained structure, where disturbance then reveals the correlations.

In condensed matter, elasticity descriptions outperform microscopic force models for almost every measurement that matters, and the same pattern also appear in physics more broadly. Effective field theories work because they describe how the medium responds, not what the ultimate constituents are. From that perspective, fields are closer to constitutive relations than to forces, as they encode how reality yields under constraint. This also explains why alternative frameworks often outperform field theory in specific domains. In fact lattice methods, holographic dualities, agent-based models, and information-theoretic approaches are most effective precisely where perturbative field theory begins to fail. What they share is a common strategy, because they behave as if the substrate carries structure forward. The vacuum then is biased, and the response depends on prior configuration rather than instantaneous interaction alone. These methods succeed precisely because they make compensations for the absence of memory as a primitive. In fact they recover behaviour that standard formulations cannot easily express by embedding history into their procedures rather than into their ontology. Therefore prediction improves when the system is treated as if it retains deformation, even though that retention is not explicitly encoded.

For example perturbative QCD works at high energies and fails at low energies. Confinement and mass gaps emerge only through lattice methods, and that behaviour matches an elastic medium that stiffens as strain accumulates.

Couplings change with scale because short-range disturbances permanently deform the background. This all appears as stress being absorbed into material properties, rather than the renormalisation of a fundamental force. In AdS/CFT correspondence bulk gravity dynamics map to boundary field deformations. The bulk behaves like a strained geometric medium whose response encodes boundary information. Forces are secondary, and stored structure does the work. Vacuum polarisation in QED is another example where the vacuum screens charge and alters propagation constants, which is textbook elastic behaviour. For effective field theories in general, low-energy parsimonious theories outperform high-energy "fundamental" ones because they describe how the substrate responds once it has already absorbed historical constraint.

From that point of view the "three" in SU(3), seen as a three branch uncertainty manifold, is the lowest dimensional structure that can host a fully antisymmetric three body state. Two branches cannot carry a baryon without violating the Pauli principle. Four or more branches generate extra states that have not been seen. The geometry of SU(3) then follows a root system at 120 degrees, a Z_3 centre, and a natural phase spacing of $2\pi/3$. DE packages this into a single statement that 'Phase is the record of how uncertainty has been resolved.' The allowed internal phases are those that let the recursion close without leaving unused branches. This would imply that there are only three stable cases with one branch, two branches, and three branches respectively. These map directly to U(1), SU(2) and SU(3), and the $2\pi/3$ angle is the geometric signature of the three branch case. The rest of this section unpacks that claim. I first review why three SU groups are needed at low energy, then explain why colour in particular needs three values, and finally show how the $2\pi/3$ phase emerges as the unique way to keep the SU(3) sector holomorphic, confining and entropy minimal.

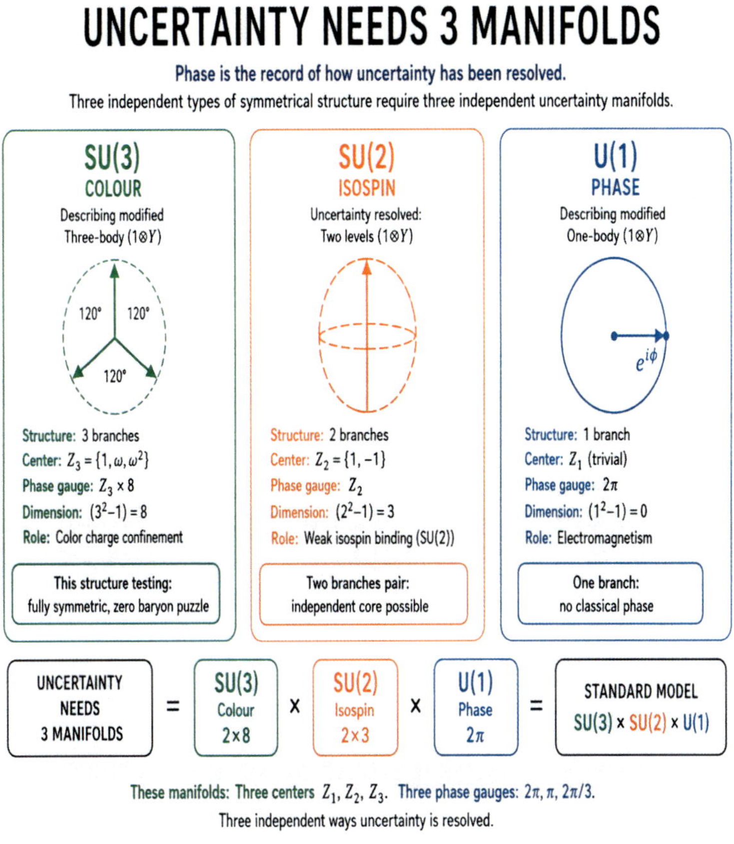

In the Standard Model three groups appear because there are three independent classes of conserved structure in low-energy physics. From a DE view, these arise because three distinct uncertainty manifolds cannot be reduced further without loss of predictive capacity. Each SU group reduces a different kind of uncertainty, SU(3) handles colour uncertainty, SU(2) handles isospin pairing, U(1) handles the residual phase.

The 3 Irreducible Uncertainties

1. Phase Uncertainty (U(1))

Is the uncertainty in the overall complex phase of the recursion. This is the uncertainty that remains even when all structural distinctions have collapsed, because information cannot be extracted by measuring global phase. It is an information-free degree of freedom, it is the uncertainty of the 1-branch manifold, where recursion closure occurs after one step.

$$N = 1, S = \ln\ 1 = 0$$

Physically this is the residual phase freedom of electromagnetism and U(1) is the symmetry that expresses the irreducible uncertainty in global phase.

2. Parity / Doublet Uncertainty (SU(2))

This is the irreducible two-state uncertainty structure that cannot be collapsed by phase redefinition, left–right, up–down, particle–partner. A 2-branch recursion cannot collapse to one branch because the recursion would oscillate:

$$\phi \rightarrow \phi + \pi \rightarrow \phi + 2\pi$$

This creates a fundamental two-state ambiguity, the uncertainty of the 2-branch manifold, which closes only after two steps.

$$\mathrm{N} = 2, \mathrm{S} = \ln\ 2$$

This gives the doublet structure of weak isospin:

- (ν, e)
- (u, d)
- left–right chirality pairing

Therefore SU(2) is the symmetry that expresses the irreducible uncertainty in binary pairing.

3. Triplet / Colour Uncertainty (SU(3))

This is the uncertainty in tri-state identity, three states that cannot be transformed into one another through binary flips or global phase. This is the minimal nontrivial antisymmetric structure. A 3-branch recursion closes only after three steps:

$$\phi \to \phi + \frac{2\pi}{3} \to \phi + \frac{4\pi}{3} \to \phi + 2\pi$$

This creates a three-way identity ambiguity, with no reduction to 1 or 2 branches preserves antisymmetry. It is the uncertainty of the 3-branch manifold, the highest stable closure.

$$N = 3, S = \ln\,3$$

This corresponds to colour uncertainty in QCD, three distinguishable colour states, three-index antisymmetric baryon singlet, centre group Z_3, and confinement from triplet closure. SU(3) is the symmetry that expresses the irreducible uncertainty in tri-state identity. These are the only uncertainties that do not collapse under recursion, that produce distinct measurable structure and preserve parsimony, and generate stable symmetries that match the Standard Model exactly. Larger uncertainty manifolds (N>3) do not appear as stable low-energy structures, and parsimony suggests that it introduces redundancy and fails to close under the DE recursion because they are outcompeted. This suggests a correspondence between irreducible uncertainty and the symmetry groups that preserve it;

Irreducible uncertainty $N \implies S = \ln N \implies SU(N)$ with the constraint:

$$N \in \{1,2,3\}$$

This provides a minimal DE rule that reproduces the Standard Model gauge group content.. To understand why nature uses the three specific gauge symmetries U(1), SU(2) and SU(3), we begin with a very simple idea. Every system carries uncertainty, and as the system evolves, it reduces this uncertainty step by step (In DE this step is termed a recursion). Imagine the system trying to correct itself after each step. For this to work, the steps must eventually close into a loop, because if the loop never closes, the system cannot predict its own behaviour. Conversely if the loop closes immediately, the system has no structure at all, so stability lies somewhere in between, and the loop can close only if the system returns to something that is recognisably the same after a certain number of steps. The number of steps it takes to close the loop tells us how many distinct internal states the system had to visit in order to get there. DE shows that only three types of loops survive once the universe settles. A one-step loop means the system has only one internal state. Nothing distinguishes one moment from the next except the overall phase of the system. This is the U(1) symmetry of electromagnetism. A two-step loop means the system alternates between two distinguishable states, and the minimal structure that produces oscillation without closure, the basic "pairing uncertainty" that underlies the SU(2) symmetry of the weak interaction. A three-step loop means the system cycles through three distinct states before closing. This is the smallest structure that creates direction and asymmetry, and it corresponds to the SU(3) symmetry of the strong interaction.

The crucial point here is that loops with more than three distinct states have not been seen and may not survive. They might either collapse into the smaller loops or destabilise under uncertainty. In other words, the universe may be pruned down to the only three stable ways that uncertainty can be optimally resolved. From this simple requirement, that uncertainty must reduce through a loop that eventually closes, the three gauge symmetries of the Standard Model fall out naturally, because they are the only stable ways that a system can carry internal distinctions without collapsing into flatness or chaos. The *i-axis* is the fundamental phase direction required for any coherent recursion. Particles are the excitations created when those stable closures exist, and this is the starting point for the more formal derivation that follows;

1. Basic DE setup

Take a state vector Ψ with an internal coordinate χ and let the system evolve by a recursion parameter ε.

- ε= recursion step (your "time" or update parameter)
- $S(\chi, \varepsilon)$= uncertainty / entropy functional
- $\Upsilon(\chi, \varepsilon)$ represents the complexity axis, defined as the accumulated constraint generated by entropy gradients across recursion. It reflects how structure

builds and persists, rather than the instantaneous gradient itself.

- the *i-axis* is the pure phase direction, so phase rotations are

$$\Psi \mapsto e^{i\theta}\Psi$$

Now define the recursion operator along the *i-axis* as

$$R(\varepsilon) \;=\; \exp\left(i\,\varepsilon\,\widehat{\Upsilon}\right)$$

where:

- ε = recursion step (ordering parameter)
- $\hat{\Upsilon}$ = generator of phase rotation determined by accumulated constraint
- *i-axis* = phase / memory axis

While entropy gradients drive change locally, Υ encodes how those gradients accumulate into stable structure across recursion, where $\widehat{\Upsilon}$ is the operator acting along the complexity axis. Gauge symmetry appears because the recursion operator must remain invariant under relabelling of phase within each closed spectrum. Its eigenvalues are

$$\lambda_k = e^{i\varepsilon\, v_k}$$

with v_k the eigenvalues of $\widehat{\Upsilon}$. So ε sets the step size on the *i-axis*, Υ sets the angular speed, and the *i-axis* is the space of pure phase rotations.

2. Closure condition on the recursion

The DE requirement is that the recursion must close after N steps for the system to be predictive:

$$R^N(\varepsilon) = \mathbb{I}.$$

In eigenvalue form this means

$$\lambda_k^N = 1 \Rightarrow e^{iN\varepsilon\, v_k} = 1 \Rightarrow N\varepsilon\, v_k \;=\; 2\pi m_k, m_k \in \mathbb{Z}.$$

So each eigenvalue corresponds to a root of unity

$$\lambda_k = e^{i\theta_k}, \theta_k = \frac{2\pi m_k}{N}.$$

The distinct θ_k give you the branches of your uncertainty manifold. Respecting DE parsimony requires the minimal N and the minimal number of distinct θ_k that still support:

- a nontrivial recursion, and
- a stable entropy partition.

This is where the 1–2–3 kernel appears.

3. Why only 1, 2, 3 branches survive

Formally, the branch count is the number of distinct eigenphases θ_k that the recursion keeps. Call this number N_b. In a maximally mixed internal sector the entropy is

$$S = \ln N_b.$$

DE demands:

1. Nontriviality: $N_b > 1$ for there to be any structure.
2. Closure: $R^N = \mathbb{I}$ for finite N.
3. Parsimony: No extra branches that do not reduce uncertainty.

Under these constraints:

- $N_b = 1$ is allowed (trivial phase freedom).
- $N_b = 2$ is allowed (binary pairing, reversible 2-cycle).
- $N_b = 3$ is the smallest irreducible oriented cycle (3-cycle that cannot be decomposed into 2-cycles without breaking antisymmetry).

For $N_b \geq 4$ we can always decompose the branch structure into smaller cycles or combinations that either:

- fail parsimony (unnecessary extra branches), or

- break closure in the presence of entropy curvature.

That is the DE reason only 1, 2, 3 are stable internal branch counts. Mathematically this says the spectrum of $\widehat{\Upsilon}$ that survives after the primordial pruning can be reduced to

$$\{1\}, \{1, -1\}, \{1, \omega, \omega^2\}$$

with $\omega = e^{2\pi i/3}$, the dynamically preferred spectrum reduces to $\{1,2,3\}$.

4. Mapping to U(1), SU(2), SU(3)

Now we match these branch structures to actual gauge groups.

4.1 U(1) from ε, Υ and the *i-axis*

For **one branch**:

- All eigenvalues are identical: $\lambda = e^{i\theta}$.
- The group of allowed transformations is all phase rotations:

$$\Psi \mapsto e^{i\alpha}\Psi, \alpha \in \mathbb{R}.$$

This is exactly U(1): the group of complex numbers of unit modulus.

In DE language:

- ε is the recursion step that causes a rotation around the *i-axis*.
- Υ acts as a scalar in this 1-dimensional internal space.
- The only uncertainty is global phase.

So the 1-branch manifold with recursion $R(\varepsilon) = e^{i\varepsilon v}$ is the U(1) sector.

4.2 SU(2) from a 2-branch recursion

Now suppose $\widehat{\Upsilon}$ has two distinct eigenvalues v_1, v_2 such that

$$e^{i\varepsilon v_1} = 1, e^{i\varepsilon v_2} = -1$$

or more generally

$$\lambda_1 = e^{i\theta}, \lambda_2 = e^{-i\theta}$$

with closure at $2\theta = 2\pi$ in the simplest case. We can represent this as an SU(2) rotation:

$$R(\varepsilon) = \exp\,(i\,\varepsilon\,\vec{n} \cdot \vec{\sigma})$$

where $\vec{\sigma}$ are the Pauli matrices and $\vec{n}$ encodes the direction of the entropy gradient in the 2-branch internal space.

Key points:

- The determinant is 1.
- The eigenvalues are complex conjugate phases on the *i-axis*.
- The internal space is 2-dimensional.

This is precisely SU(2). The 2-branch uncertainty corresponds to a doublet, i.e. a weak isospin pair.

So:

- ε = is the recursion step that causes a rotation around the *i-axi*s.
- ϒ = vector in SU(2) (via $\vec{n} \cdot \vec{\sigma}$)
- recursion $R(\varepsilon) = \exp(i\varepsilon\hat{\Upsilon})$ generates SU(2) rotations.

4.3 SU(3) from a 3-branch recursion

For three branches, take three eigenvalues

$$\lambda_1 = 1, \lambda_2 = e^{2\pi i/3}, \lambda_3 = e^{4\pi i/3}.$$

These are the cubic roots of unity, all lying on the *i-axis* circle, spaced by $2\pi/3$. The minimal traceless generator with these eigenvalues can be written as a diagonal SU(3) element:

$$\widehat{\Upsilon} = \text{diag}(0, \frac{2\pi}{3\varepsilon}, \frac{4\pi}{3\varepsilon})$$

up to an overall trace removal and basis choice. More generally, any SU(3) element is:

$$R(\varepsilon) = \exp\,(i\,\varepsilon \sum_{a=1}^{8} v_a \lambda_a)$$

with λ_a the Gell-Mann matrices and v_a components of the entropy-gradient in the 8-dimensional adjoint space.
The important DE facts:

- The eigenvalues of a generic SU(3) element are three phases whose product is 1.
- The simplest nontrivial cycle is exactly the 3-cycle

$$\phi \;\rightarrow\; \phi + \frac{2\pi}{3} \;\rightarrow\; \phi + \frac{4\pi}{3} \;\rightarrow\; \phi + 2\pi.$$

- This 3-cycle is the tri-state identity uncertainty.

So:

- ε is the recursion step that causes a rotation around the *i-axis.*
- Υ is now an element of SU(3) that picks out the three-branch spectrum.
- The recursion operator $R(\varepsilon) = \exp\big(i\varepsilon\widehat{\Upsilon}\big)$ generates SU(3) colour rotations.

5. In one compact statement

Starting from:

$$R(\varepsilon) = \exp\,(i\varepsilon\widehat{\Upsilon})$$

acting as a pure phase rotation along the i-axis, with closure condition

$$R^N(\varepsilon) = \mathbb{I},$$

DE shows:

- 1 distinct eigenphase → 1-branch manifold → U(1)
- 2 conjugate eigenphases → 2-branch manifold → SU(2)
- 3 cubic-root eigenphases → 3-branch manifold → SU(3)

and higher branch counts are pruned dynamically by parsimony and instability of closure. That is the formal DE derivation path:

$$\varepsilon, \Upsilon, i\text{-axis} \Rightarrow \text{closure spectrum} \Rightarrow \{U(1), SU(2), SU(3)\}.$$

Without distinct uncertainty manifolds, the entropy field becomes flat with substantially reduced gradients. If a universe had only global phase uncertainty (U1), no binary pairing uncertainty (SU2), and no tri-state identity uncertainty (SU3), then the recursion would collapse in one step:

$$N = 1 \Rightarrow S = 0.$$

This produces no sustained structure, no stable forces, and no persistent dynamics. A universe with only U1 would fall to pure informational equilibrium. The SU2 and SU3 manifolds prevent this collapse and SU(2) 2-branch uncertainty introduces the first nontrivial gradient:

$$S = \ln 2.$$

Binary pairing creates broken symmetry, chirality, weak interactions, matter–partner asymmetry, and early-universe phase transitions. This prevents collapse into a trivial phase-only state. SU(3) 3-branch uncertainty creates the strongest gradient:

$$S = \ln 3.$$

Tri-state identity produces confinement, stable baryons, and long-term structure. This prevents thermodynamic and informational collapse. SU(3) therefore acts as the load-bearing manifold. Without it, no persistent composite structure can exist.

The particles are the embodiment of these uncertainty manifolds, the physical excitations of these irreducible uncertainty classes:

- photons: residual-phase uncertainty (U1)
- W/Z bosons: binary-pairing uncertainty (SU2)
- gluons: tri-state identity uncertainty (SU3)
- quarks: carriers of the SU3 uncertainty itself
- leptons: minimal carriers of SU(2) pairing structure

These particles are the physical expression of the irreducible uncertainty manifolds. Removing them would collapse the corresponding entropy gradients, and without those gradients, no scaling structure can persist. Stable structure requires non-flat uncertainty and recursive closure, which occurs only for N = 2 or 3.

If we want to understand why the universe settled on the specific forces we see today, we usually begin with the mathematics of gauge theory, yet gauge theory does not explain why those symmetries are selected. DE takes a different path which begins with the more primitive idea that every system carries uncertainty and must reduce that uncertainty in a stable way. When the early universe was young, almost every possible arrangement of uncertainty collapsed, but only a few structures survived. Those survivors became the rules that shape everything. What is remarkable is that when we follow this logic through, the three fundamental forces in the Standard Model of electromagnetism, the weak interaction, and the strong interaction then fall out naturally. They are the only three patterns of uncertainty stable enough to survive the early informational storm.

If this picture is correct, no additional gauge structures are required to sustain low-energy physics. The 1–2–3 kernel provides the minimal architecture capable of maintaining structure. In the early universe, many patterns of uncertainty were explored, but only a small subset survived the pressure of rapid change and extreme gradients. What remained were the configurations capable of stable closure, and these define the rules that govern all subsequent structure.

Together they lock in the conditions that allow particles to exist, forces to stabilise, structures to form, time to unfold and let complexity to grow, the scaffolding that holds the universe open. Each constant stabilises a specific aspect of the uncertainty landscape, preventing collapse along a particular structural axis.

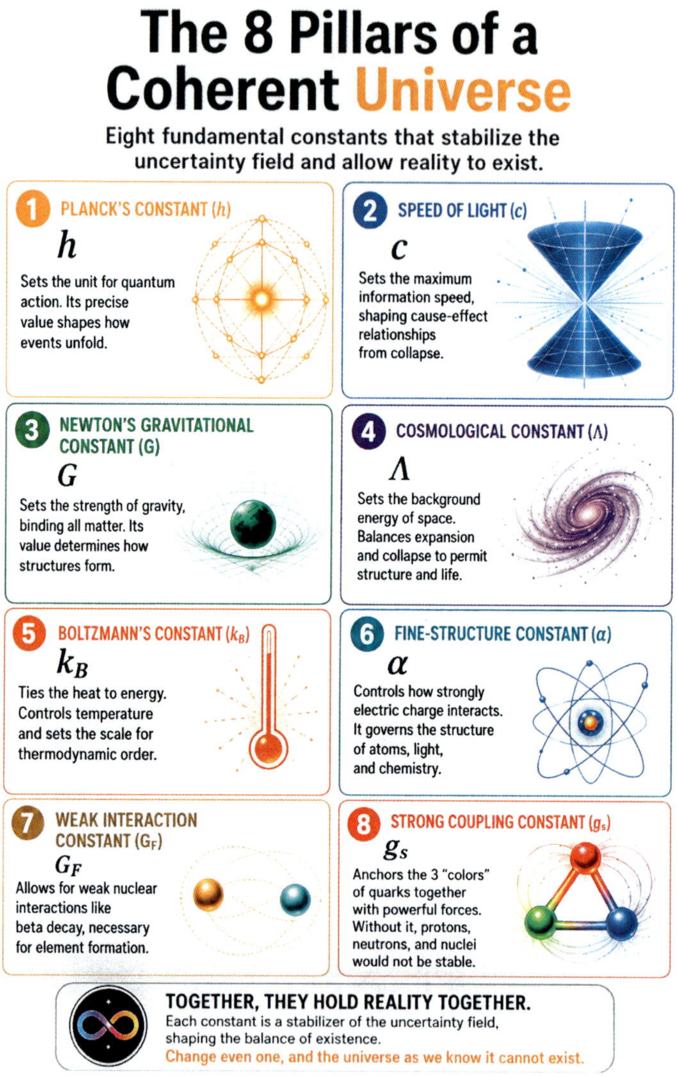

These constants are the residual constraints left after the primordial pruning of uncertainty. Each preserves a specific structural condition required to prevent collapse, and together they stabilise the 1–2–3 backbone on which all higher structure depends.

Colour

When people hear the word "colour" in physics, it is often introduced as a charge carried by quarks and mediated by gluons. In the DE framework, colour is more fundamental, as it is the minimal structure required to preserve identity within a triadic uncertainty manifold. Once phase (U(1)) and pairing (SU(2)) are established, a third constraint becomes unavoidable. Two states can oscillate and exchange structure, but they cannot stabilise a composite object without losing identity under exchange. A third branch introduces a closed cycle that preserves antisymmetry. This is the smallest structure in which distinct identities can coexist without collapsing into equivalence. This closure appears as a three-step phase cycle:

$$\phi \to \phi + \frac{2\pi}{3} \to \phi + \frac{4\pi}{3} \to \phi + 2\pi$$

Each step represents a distinct internal state, and only after the full cycle does the system return to itself. This is the origin of the three colour degrees of freedom. They are the minimal set required for closure under recursion while preserving identity. In this sense, colour encodes triadic identity, and it ensures that composite states, such as baryons, can exist without violating antisymmetry. The SU(3) symmetry then arises as the natural way to transform between these three states while preserving the closure condition. Confinement follows directly, because the triadic structure must remain intact. Isolated colour cannot exist without breaking closure, so observable states must be colour-neutral combinations. Phase defines existence, pairing defines relation, and colour defines identity.

Together they form the minimal architecture required for persistent structure.

Flavour

When people hear the word "flavour" in physics, it often appears as an additional layer of structure. Why should electrons have heavier counterparts such as the muon and tau? Why should quarks come in three families?
In the DE framework, flavour appears only after the primary uncertainty structures U(1), SU(2), and SU(3) are established. These define phase, pairing, and identity, so that once these are fixed, flavour emerges as a constrained degree of variation within that existing structure.

For example two states may share identical charge, colour, and pairing properties, yet differ in how they traverse the same uncertainty manifold. This difference appears as variation in effective recursion rate, which manifests physically as mass and interaction scale. Flavour therefore parameterises inequivalent trajectories within the same SU(2) × SU(3) framework. The existence of three generations follows from the same closure constraint already identified. The SU(2) × SU(3) backbone admits a minimal set of distinct, stable trajectories before redundancy appears. Fewer would collapse distinctions between states, while more would introduce unstable or unobserved configurations. In this sense, flavour is a constrained variation within a fixed symmetry structure. Colour defines identity, weak isospin defines pairing, and flavour distinguishes how otherwise identical states evolve within that structure. It is not an additional degree of freedom, but a limited modulation of the existing uncertainty manifold.

The Power of Three: Colours, Particles, Codons

Certain patterns in nature repeat often enough that they demand explanation. Three quark colours, three generations of particles, three bases per codon in DNA. These are reflections of a deeper constraint on how structure can stabilise. In DE, stability requires that uncertainty closes into a loop. A single-step loop produces no structure, a two-step loop oscillates without direction, but a three-step loop is the first configuration that supports identity, orientation, and stable return. This minimal cycle corresponds to a phase rotation of: $0 \to 2\pi/3 \to 4\pi/3 \to 2\pi$,
and is the smallest non-trivial closure that avoids collapse.
In quantum chromodynamics, this appears as SU(3). Its root system is spaced at 120 degrees, its centre is Z_3, and its fundamental representation carries three distinct phase states. The same structure emerges from recursion, three steps are required to return to identity without degenerating into symmetry or instability. The $2\pi/3$ phase is therefore the geometric signature of triadic closure.
This becomes physically necessary when we consider fermions. For the Δ^{++} baryon (uuu), all three quarks share the same flavour, spin, and spatial state. Without an additional internal label, the total wavefunction would be symmetric, violating the Pauli principle. The solution is a three-valued internal degree of freedom, colour, which allows a fully antisymmetric combination:

$$1/\sqrt{6}\,(rgb + gbr + brg - rbg - grb - bgr).$$

This is the unique antisymmetric 3-index structure, the minimal number that satisfies antisymmetry, confinement, and the observed hadron spectrum. The underlying algebra is captured by the cubic roots of unity:

$$\omega = e^{\{i2\pi/3\}},$$

$$\omega^3 = 1,$$
$$1 + \omega + \omega^2 = 0.$$

These define a Z_3 phase grading. Any state decomposes into three eigenphases, and three such phases sum to zero, allowing closure into a neutral object. This directly produces:

- three colour directions (eigenstates of phase),
- three-body colour neutrality (baryons),
- the antisymmetric singlet structure.

In QCD, this appears through the centre group:

$$Z_3 = \{1, \omega, \omega^2\}.$$

A Wilson loop may acquire one of these phases, and confinement follows from the requirement that only combinations whose total phase multiplies to unity are observable. This forces:

- $three\ quarks \rightarrow \omega \cdot \omega \cdot \omega = 1,$
- $quark - antiquark \rightarrow \omega \cdot \omega^2 = 1.$

The closure rule is therefore fixed by phase, not imposed externally. The same structure appears in recursion. A phase step of $2\pi/3$ produces three distinct states before closure:

$$1 \rightarrow e^{\{i2\pi/3\}} \rightarrow e^{\{i4\pi/3\}} \rightarrow 1.$$

This yields three linked levels:

1. **Branch structure** three distinct internal states
2. **Closure structure** minimal three-step orbit
3. **Entropy structure** balanced partition:

$$1 + \omega + \omega^2 = 0$$

This is the minimal non-trivial entropy configuration that cannot be reduced to a dyad without breaking symmetry. In compact form:

$$2\pi/3 \rightarrow \omega$$
$$\omega \rightarrow Z_3\ grading$$

$$Z_3 \rightarrow colour, confinement, and\ three - branch\ recursion$$

The same structure appears at the algebraic, physical, and dynamical levels.

Why six quarks

The existence of six quarks follows from three independent constraints:

- SU(2) pairing (weak isospin doublets)
- SU(3) tri-state identity (colour)
- anomaly cancellation and CP violation

These combine to produce a minimal consistent structure. Each quark must exist in a left-handed SU(2) doublet:

$$(u, d), (c, s), (t, b).$$

Anomaly cancellation requires that each lepton generation is matched by a quark doublet with three colours. This fixes the number of generations to three, where CP violation provides an additional constraint. With two generations, all complex phases in the CKM matrix can be removed. With three, an irreducible phase remains, which is observed experimentally. This confirms that the system cannot terminate at four quarks and mass hierarchy stability further restricts the count. With fewer generations, the structure maybe underdetermined, yet with more, flavour-changing processes could become unstable and conflict with observation. From the DE perspective, the same result follows directly. Colour provides three branches, weak isospin provides two states per branch, and recursion closure produces a 3×2 grid of allowed states. This yields six quarks as the minimal stable configuration, and any attempt to add or remove states breaks closure, destroys anomaly cancellation, or eliminates CP violation. So six is the smallest structure that satisfies all constraints simultaneously.

Transition: flavour, mixing, and time asymmetry

Within this structure, flavour does not introduce a new symmetry. It parameterises inequivalent trajectories within the SU(2) × SU(3) manifold. States with identical gauge structure can differ in how they traverse the uncertainty landscape, appearing as differences in mass and interaction scale. The crucial consequence is that mixing becomes possible. When three generations are present, the transformation between flavour states cannot be reduced to a simple rotation. Instead, it acquires an irreducible complex phase, encoded in the CKM matrix. This phase cannot be removed by any redefinition of fields. It represents a structural asymmetry in how states evolve, and it is the origin of CP violation. In DE terms, this arises because a three-branch system supports orientation in a way a two-branch system cannot. A triad introduces ordering that cannot be undone without breaking closure, thereby creating a preferred direction in the recursion. CP violation is therefore a direct consequence of this triadic structure. Once three branches are present, phase cannot be globally eliminated, and asymmetry becomes unavoidable. This connects the algebra of SU(3) directly to temporal structure. The same three-branch closure that stabilises colour also permits irreducible phase in flavour mixing. From this point of view, time asymmetry is already encoded within the minimal structure required for stable matter.

Any SU(N) group has roots spaced by $2\pi/N$. For N=3 it gives the familiar threefold symmetry. Twistor geometry shows the same spacing when colour is embedded as a projective phase, so SU(3) becomes inevitable. For a gauge force that confines by a threefold phase winding and returns to identity after three rotations then SU(3) is the smallest group that fits. Anything smaller will not confine and anything larger adds structure that is not observed. So three SU sectors and three couplings arise because the physical world separates its uncertainty surfaces into exactly three independent rotations. The same threefold structure appears across multiple domains, because it is the minimal configuration that supports closure without collapse. In algebra it appears as the cubic roots of unity, in gauge theory as the SU(3) symmetry and its Z_3 centre, and in particle structure as colour confinement and three-body antisymmetry. These are not separate phenomena but different expressions of the same constraint, where three is the smallest structure that preserves identity, orientation, and return. Once this condition is satisfied, the details vary, but the underlying form remains fixed.

Branch Closure and the Uniqueness of 1–2–3

To establish the Standard Model structure it is not sufficient to show that 1-, 2-, and 3-branch recursion is viable, we must also show that higher branch counts do not survive under the same constraints. Let the recursion operator act on an internal phase space with eigenvalues:

$$\lambda_k = e^{2\pi i k/N}, k = 0,1,\dots,N-1$$

Closure requires:

$$R^N = \mathbb{I}$$

so that recursion returns to identity.

1: Minimal Closure

A viable recursion must return to identity do so with minimal phase redundancy:

- $N = 1$: trivial closure (U(1))
- $N = 2$: reversible dyadic oscillation (SU(2))
- $N = 3$: minimal nontrivial cycle with orientation (SU(3))

For $N > 3$:

- closure occurs but introduces redundant phase states

2: Parsimony (DE selection rule)

Define entropy of internal branch structure:

$$S_N = \log N$$

A system is viable only if additional branches reduce uncertainty per recursion step:

$$\frac{\Delta S}{\Delta \varepsilon} < \text{resolution capacity}$$

For $N > 3$:

- entropy increases faster than structure
- additional branches do not improve predictive compression

Thus higher N increases cost without increasing resolution.

3: Stability Under Composition

Consider combining subsystems.

For $N = 2$:

- system remains reducible (pairwise symmetry)

For $N = 3$:

- first irreducible antisymmetric structure appears:

$$\epsilon_{ijk}$$

For $N > 3$:

- antisymmetric tensors exist but introduce degenerate or unobserved states. Thus $N = 3$ is the smallest structure supporting stable antisymmetry without redundancy.

4: Recursive Viability

Let recursion reduce accessible state space:

$$\Omega_{n+1} = \alpha^{-1}\Omega_n$$

For stability:

$$N \leq \frac{1}{1 - \alpha^{-1}}$$

With:

$$\alpha^{-1} \approx 0.4$$

we obtain:

$$N \lesssim 3$$

Thus higher branch counts exceed viable compression capacity. Only $N = 1,2,3$ simultaneously satisfy closure, parsimony, antisymmetry, and recursive stability. Hence this yields the unique structure:

$$U(1) \times SU(2) \times SU(3)$$

This result constrains the minimal structure required for stable recursion. In conventional physics gauge groups are postulated and their origin is unexplained, but in this frame gauge groups emerge as the only stable branch closures as their dimensionality is fixed by recursion constraints. Thus the Standard Model becomes the minimal stable solution to recursive entropy resolution as higher number of N introduce redundant degrees of freedom that do not improve predictive compression, and are therefore pruned under DE recursion.

CKM Phase as Triadic Holonomy

Within the Standard Model flavour mixing is encoded in the CKM matrix. Empirically, this matrix contains an irreducible complex phase responsible for CP violation. The origin of this phase is usually treated as a structural feature of the theory, but within DE it follows directly from the transition from dyadic to triadic closure. To see this, consider first a system with only two generations. Any mixing between two states can be represented as a rotation. Complex phases can always be absorbed into redefinitions of the two fermion fields, leaving a purely real transformation. The system supports exchange, but not orientation. In this sense, two-state mixing is reversible because no physically meaningful phase remains.

The situation changes qualitatively when a third generation is introduced. With three coupled states, mixing is no longer reducible to pairwise exchange as the system now contains a closed loop of transitions. While local phase redefinitions can remove phases along individual links, they cannot eliminate the phase accumulated around the loop as a whole. A residual phase remains, and this phase is observable. This can be expressed in minimal form by considering three transition amplitudes between states 1, 2, and 3:

$$U_{12}, U_{23}, U_{31}.$$

The phase accumulated around the loop is:

$$\Phi_{\Delta} = \arg\left(U_{12}U_{23}U_{31}\right).$$

This quantity is invariant under rephasing of the individual states. In a two-state system, no such non-trivial loop exists, and the corresponding phase vanishes. In a three-state system, the loop is unavoidable, and so the phase survives. This is the minimal algebraic expression of CP violation.

In geometric terms, a dyad defines a line segment, while a triad defines a closed triangle. Along a line, phase can be cancelled by reversal, but around a loop phase can accumulate. The residual phase is therefore a holonomy, a measure of how much the system fails to return to its original state after transport around a closed path, and within DE, this distinction is fundamental. Once three branches are coupled, recursion no longer closes trivially, it closes with a residue, and the CKM phase is this residue. It is the first non-trivial consequence of triadic structure. The same three-branch closure that stabilises colour also permits an irreducible phase in flavour mixing. From this perspective, CP violation is the algebraic signature of triadic closure. The asymmetry it introduces reflects a deeper structural fact: once a system contains three coupled states, phase cannot be globally eliminated, and directionality becomes unavoidable. In this way, the origin of time asymmetry is already encoded within the minimal structure required for stable matter.

Colour, Confinement, and Direction

Colour therefore is a constraint on how information can be distributed within a strongly interacting system. In the SU(3) sector, three distinguishable states form a closed phase structure:

$$1, \omega, \omega^2 \text{ with } 1 + \omega + \omega^2 = 0,$$

where $\omega = e^{i2\pi/3}$. This identity enforces balance. Three branches can close into a neutral configuration, while any subset cannot. Confinement follows directly, as only combinations whose total phase returns to unity are observable. A baryon requires three quarks, and a meson requires a quark–antiquark pair, because these are the minimal combinations that satisfy the closure condition. From an informational perspective, colour is a three-state variable with entropy

$$S = \ln\ 3.$$

A free quark would carry this entropy across macroscopic scales, increasing the information cost. Confinement reduces that cost by forcing colour-neutral combinations, compressing the entropy of three quarks into a single observable state:

$$\Delta S = -\ln\ 3.$$

The strong interaction therefore enforces minimal informational structure. Colour is a closed memory system and information circulates internally but cannot be exported without breaking closure. This same three-branch structure also introduces direction. A single branch has no ordering, and a two-branch system oscillates reversibly. Only a three-branch cycle produces a sequence that cannot be undone:

$$1 \to \omega \to \omega^2 \to 1.$$

Forward and backward traversals no longer coincide. The mismatch between them defines orientation. Algebraically, this appears as the antisymmetric tensor ε_{abc}, whose sign flips under exchange. That sign is the minimal expression of an arrow. Therefore in DE terms, time emerges when recursion proceeds on a triadic manifold that cannot be reduced to a reversible cycle. Entropy increases because the system follows an oriented sequence, not the other way around. The same structure that enforces confinement therefore also introduces direction. Colour stabilises space by closing information, while triadic recursion introduces ordering that cannot be reversed. Together, these form the minimal conditions for persistent structure. A system must be able to retain identity, enforce closure, and sustain directional recursion. The SU(3) sector provides all three, and this is why the strong interaction is not only responsible for binding matter, but for maintaining the conditions under which structure can persist across scales.

At this point ε becomes monotonic. Each step along the cycle reduces unresolved uncertainty while preserving a trace of what has been excluded. The system no longer revisits prior states in a fully reversible way. Instead, it accumulates ordered structure, and this accumulation is what defines entropic distance. Formally, ε measures the progressive contraction of accessible microstates. At each step, the system eliminates configurations that are no longer compatible with its constraints. The remaining space shrinks, but the information required to describe the system increases. Resolution is multiplicative, but memory is additive. This establishes a key relation:

- ε orders resolution
- phase generates recursion
- closure enforces selection
- memory records what survives

In this frame, ε is the depth of resolved uncertainty, and time appears only when ε becomes directional, when recursion cannot be undone without violating closure. The triadic structure is therefore the minimal condition under which ε acquires structural meaning. Only then can a system move away from its past in a way that cannot be retraced without cost. This provides the bridge to collapse. As ε increases, the system approaches a point where maintaining multiple possibilities becomes more expensive than selecting one.

At that threshold, recursion can no longer sustain ambiguity, and the system commits to a single branch. Collapse occurs when entropic distance exceeds the capacity to hold unresolved states. In this way, confinement, direction, and collapse are not separate phenomena. They are different expressions of the same underlying process, that is the evolution of a system along ε under triadic constraint. Colour enforces closure, recursion generates sequence, and ε measures the depth of that sequence as it becomes irreversible.

The Standard Model is the minimal structure required for a universe that must resolve uncertainty while preserving constraint and continuity. Phase defines what can exist, pairing defines how states relate, and triadic identity defines how they persist. From these, symmetry, particles, and interactions follow as direct consequences. Colour enforces closure, flavour encodes variation, and the residual phase fixes orientation. Together they generate a system that can stabilise structure while continuing to evolve. Once this architecture is in place, direction emerges automatically.

Triadic closure introduces an irreducible phase, that phase accumulates along recursion, and ε becomes monotonic. Time, asymmetry, and collapse are the unavoidable outcome of a system that can no longer sustain multiple possibilities at once. What appears as measurement, confinement, or CP violation are different faces of the same transition, the point at which preserving possibility costs more than selecting a path. In this sense, the Standard Model is the first stable solution to recursive entropy resolution. It is the simplest structure that can carry memory, enforce constraint, and allow complexity to grow without collapsing into symmetry or dispersing into noise. Everything that follows is built on that foundation.

THE DE 1-2-3 FRAMEWORK

TEN BRIDGES THAT FIX THE SU(3) FOUNDATION

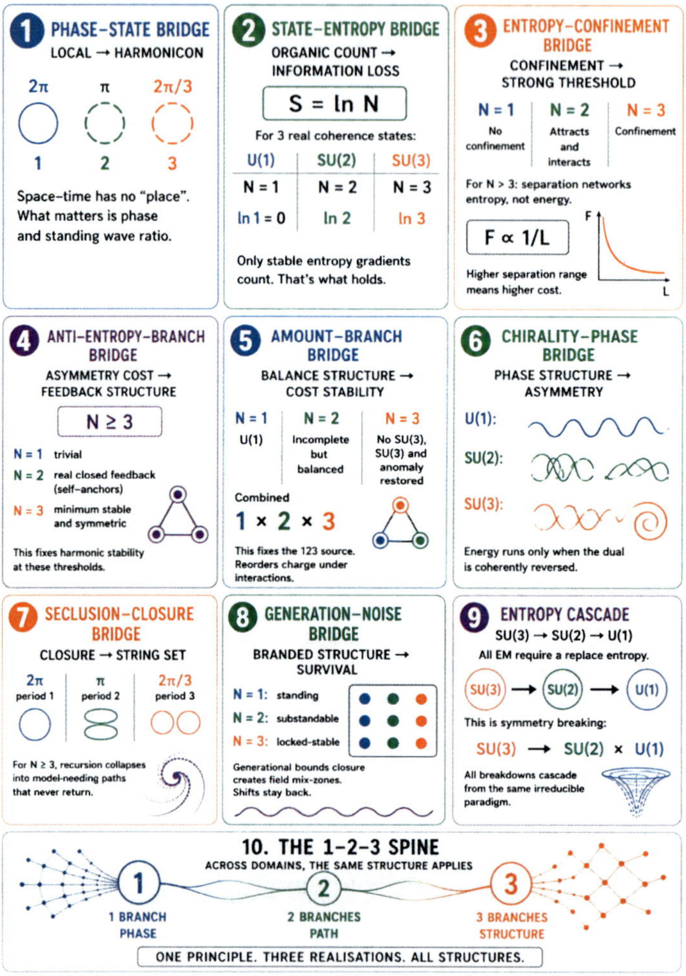

Chapter 4: The Complexity Kernel

In the last chapter we established that only one-, two-, and three-branch uncertainty manifolds survive Deductive Entropy pruning, and that these map uniquely to U(1), SU(2), and SU(3). The 1–2–3 pattern is not domain-specific, it is the minimal configuration capable of sustaining identity, contrast, and closure simultaneously. Wherever structure persists, wherever direction appears, and wherever complexity accumulates, this same backbone re-emerges as a projection into each different domain.

What matters therefore is how these symmetries propagate. The same closure conditions that stabilise phase, pairing, and triadic identity in particle physics also govern the formation of structure at every scale. Intuitively it can be said that the universe does not switch rules as it grows more complex, it reuses the same minimal architecture under different constraints. Starting from phase, only three closure patterns survive, as a one-step return, a two-step oscillation, and a three-step cycle that preserves orientation. These generate the 1–2–3 backbone that reappears at every level of description. Phase angles fix the geometry, branch count fixes the informational degrees of freedom, entropy follows as $S = \ln N$, and symmetry emerges as the minimal algebra required to preserve those structures.

The 1–2–3 DE Complexity Kernel

What follows is a catalogue of how that fixed structure projects into every domain where stability, direction, and complexity appear. The 1–2–3 structure can be expressed as a layered projection from phase to physical law, where each layer follows directly from the one before it. The DE attractor structure selects exactly N = 1, 2, 3 as the only DE-stable attractors.

This sequence is enforced by closure as the smallest structure capable of sustaining recursive stability. No additional low-energy branch structure survives DE pruning, because excess degrees of freedom introduce entropy that cannot be closed. A stable system cannot support more informational branches than it can return to identity. For three branches, the minimal state is a three-component complex vector. Preservation of total probability requires invariance of the norm, restricting transformations to unitary operations. Removing the redundant global phase imposes det = 1, eliminating unphysical scaling. These constraints uniquely select SU(3) as the symmetry that preserves triadic closure. The dimensionality follows from the number of independent generators:

$$\dim SU(N) = N^2 - 1 \ \Rightarrow \ \dim SU(3) = 8$$

These eight generators, represented by the Gell-Mann matrices, define the independent transformations of colour charge and give rise to the gluon octet. This is the minimal structure: smaller groups cannot rotate three independent states, while larger groups introduce redundant degrees of freedom that are not realised. The strong interaction follows directly from this structure. SU(2) supports reversible pairing between two states. SU(3) is the first structure that supports non-reversible circulation across three states, and therefore stable identity. This arises because stable recursion can close only at angles producing integer branch counts. The layered structure that follows is the direct projection of this closure constraint from phase into physical law. All stable structure therefore is the consequence of a single constraint, closure without collapse exists only for N = 1, 2, and 3.

THE 1-2-3 STRUCTURE OF NATURE

DERIVED FROM PHASE GEOMETRY, INFORMATION, AND DE RECURSION

	U1 (N+1)	SU2 (N+2)	SU3 (N+3)	
LAYER 1 PHASE AXIUS (GEOMETRY) Two-π or top-two stable phase (nπ:1:50).	2π One-π period = 1	π One-π period = 2	$\frac{2\pi}{3}$ One-π period = 3	GEOMETRIC FUNDAMENTALS
LAYER 2 BRANCH COUNT (INFORMATION TOPOLOGY) Judge from the number of allowed branches.	1 branch	2 branches	3 branches	1–2–3 BRANCH KERNEL
LAYER 3 INFORMATIONAL ENTROPY For a maximally parsed N-state array with source (dim index S = ln K	$S = \ln 1 = 0$	$S = \ln 2$	$S = \ln 3$	ENTROPY SPINE
LAYER 4 STABILITY STRUCTURE (LIFE CODING) Stable down coding rule in life cycle waves (E-jam state).	1 branch → U(1) Identity/phase group (Electromagnetism)	2 branches ↔ SU(2) Special unitary group (Weak interaction)	3 branches ↔ SU(3) Special unitary group (Strong interaction)	THE STANDARD MODEL GAUGE GROUPS
LAYER 5 THERMODYNAMIC HARMONICS (FORCE STRUCTURE) Force loop and avalanche of thermodynamic behavior.	1 branch: no coherence, zero phase	2 branches: standing wave structure	3 branches: coherence because superposition resonance energy core	THERMODYNAMIC BEHAVIOR EXPLANATION TO FORCE
LAYER 6 ANTISYMMETRY AND FORMATION STRUCTURE Force-based order causes multi-ramble compaction and spiral.	1 branch: no net polarity linear array	2 branches: axis of return Hook and geometric star	3 branches: triiral filling and geometric 3-lock star coding	ALLIED FORCES TO FORM AND PRESS CLOSURE TO 3
LAYER 7 ANOMALY STRUCTURE AND GENERATION Anomalous decay and generation are 1-2-3 real universal cascade in source force and geometric EMI Jam all.	1-2-3 = 6 axial clapses per generation	Anomalies cancel with 2 parallels XXY lock at mid-branch 2	No triple-N breach structure is consistent with local symmetry	ANOMALY CANCELLATION AND MATTER CONSISTENCY
LAYER 8 CP VIOLATION AND CHIRALITY Opposite parity tilt, source-to-center to nπ:3 T/L J ratio.	1 branch: no mixing	2 branches: right and left CP Polatization	3 branches: CPV for mis-triangle, real CP violation	CP VIOLATION FROM 2, STRUCTURE
LAYER 9 DE RECURSION AND CLOSURE Cascade to next return to base 1-2-3 reach predictive.	1 cycle → unwilled seed	2 cycle ↔ parity enclosure	3 cycle ↔ tririal wound structure	DE RECURSIVE RETURN TO 1 → 1-2-3 ATTRACTOR SEQUENCE
	> 3	>> 4 — toradic recursion with reccur attractivity		HIGHER FORCE OF 16 PATHWAYS

The diagram below shows the structure in its simplest form. From this, the behaviour of complex systems follows directly. Once more than one path exists, the system must organise how those paths relate. With two possibilities, motion is reversible, the system oscillates and retraces its steps, so no structure accumulates. With three possibilities, exact retracing is no longer possible. The system must circulate, and that circulation introduces orientation, memory, and stable identity. This is the first point at which structure can persist and propagate.

SU(2) Dyad

SU(3) Triad

The distinction is geometric. A two-branch system forms an oscillation, while a three-branch system forms a closed loop. Closure introduces direction, and direction allows memory to accumulate. This defines the branch-parsimony ladder:

· $2\pi \rightarrow 1$ path
· $\pi \rightarrow 2$ paths
· $2\pi/3 \rightarrow 3$ paths
Only these phase rotations produce stable recursive closure. Any other angle either fails to close or generates excess entropy that cannot be stabilised. From this constraint, the Standard Model gauge groups follow as the only symmetries compatible with stable closure:
· U(1): single-phase freedom
· SU(2): reversible pairing
· SU(3): triadic closure with confinement

These are the only organisational states that allow uncertainty to resolve without loss of structure. U(1) describes pure phase rotation and is therefore Abelian. SU(2) is the minimal non-Abelian group, supporting reversible mixing between two states and giving rise to the weak interaction. SU(3) is the first structure that supports non-reversible circulation across three states. Its eight generators define the independent transformations of colour charge, producing gluons, confinement, and stable baryonic structure. The hierarchy follows directly from the entropy-angle bridge, where the number of viable paths scales with the allowed closure angle, and the informational cost scales with the number of paths. Stable structure therefore exists only at the angles that permit recursive closure.

Bifurcation as Viability Threshold

Once a complex-conjugate pair appears, time-like evolution emerges as phase winding between the two components, and that winding is oscillatory by construction. Complex eigenvalues come in pairs, and paired degrees of freedom with conserved norm necessarily imply oscillation, which is the minimal behaviour compatible with a dyad that preserves identity under recursion. Only later, when additional structure arrives, does oscillation become directional, damped, frozen, or locked into mass-bearing modes. The transition from oscillatory uncertainty to committed structure is selected by viability.

1. Collapse as a comparison of costs

In DE, two entropic costs coexist:

A. The cost of maintaining microstate uncertainty.

Let ε measure unresolved uncertainty. The cost of holding an undecided distribution is accumulated along the entropic coordinate:

$$W_{\text{hold}}(\varepsilon) = \int_{\varepsilon_*}^{\varepsilon} \Phi(u)\, du,$$

where $\Phi(\varepsilon)$ is the marginal cost of keeping alternatives simultaneously viable. This cost reflects coupling strength, constraint pressure, and the increasing fragility of broad superposition.

B. The cost of committing to a macrostate.

Committing requires writing and stabilising memory:

$$W_{\text{commit}}(\varepsilon) = W_{\text{write}} + W_{\text{stabilise}}.$$

This is the cost of selecting a branch and making that selection robust.

2. The true collapse condition

Collapse occurs not because entropy "pushes" the system, but because remaining undecided becomes more expensive than choosing.

$$\boxed{W_{\text{hold}}(\varepsilon) > W_{\text{commit}}(\varepsilon)}$$

At this threshold, ambiguity is no longer viable. Commitment becomes the cheaper configuration, and this is the Deductive Entropy trigger. The system collapses because persistence becomes less costly than continued uncertainty.

3. The universal contraction enters

Near such viability thresholds, many nonlinear systems exhibit renormalisation-like behaviour, and unresolved uncertainty contracts by a near-constant ratio per step. Let

$$\varepsilon_{n+1} = \frac{\varepsilon_n}{\alpha_{\text{eff}}}, \alpha_{\text{eff}} > 1.$$

This reflects the universal geometry of coarse-grained contraction near bifurcation. The logarithmic form is key:

$$\Delta \ln \varepsilon = -\ln \alpha_{\text{eff}}.$$

Resolution proceeds multiplicatively, but memory, however, accumulates additively:

$$\Delta \Upsilon = \ln \left(\frac{\varepsilon_n}{\varepsilon_{n+1}}\right) = \ln \alpha_{\text{eff}}.$$

This is the complexity ratchet, where each contraction of uncertainty produces a fixed increment of locked history.

4. The retention fraction

If a collapse "click" retains a fraction f of the previous accessible uncertainty, then

$$f = \frac{\varepsilon_{n+1}}{\varepsilon_n} = \frac{1}{\alpha_{\text{eff}}}.$$

For $\alpha_{\text{eff}} \approx 2.5$,

$$f \approx 0.40.$$

So each step retains roughly forty percent of prior freedom and permanently sequesters the rest into structure. This is a compression ratio, and the logarithmic increment in complexity per step is:

$$\Delta\Upsilon = \ln\, \alpha_{\text{eff}}.$$

The Feigenbaum constant describes the geometry of contraction when bifurcation becomes viable. Thermodynamics supplies bounds on physical implementations of memory writing, but it is not the driver. The driver is entropic viability, the crossing point where maintaining uncertainty costs more than committing to a structured state. In this framing oscillation is the minimal dyadic behaviour, bifurcation occurs when ambiguity becomes unstable, so collapse is chosen because it is cheaper. Resolution is multiplicative, memory is additive, and collapse is economic.

At each viability threshold, contraction selects which portion of phase survives. Phase may evolve freely, but only the component compatible with decreasing uncertainty is retained. This retained component forms the complexity increment, so that memory is the portion of phase that passes through the contraction filter.

From Oscillation to Time

In the pure dyadic regime, phase ϕ advances but no irreversible structure accumulates:

$$\Upsilon = 0$$

The system remains reversible and no history is written. Once contraction begins and $\Upsilon > 0$, phase advance is no longer reversible. Ordering becomes cumulative and memory enforces asymmetry. Time becomes directional precisely when phase survives contraction. The contraction ratio introduces structure, and:

$$\Upsilon \sim \frac{i}{\alpha_{\text{eff}}}$$

captures the minimal rule that complexity is renormalised phase memory.

1. Collapse as Entropic Viability

Let ε denote entropic distance, the remaining unresolved uncertainty. Two competing costs govern the system:

- **Delay cost**: maintaining uncertainty
- **Commitment cost**: writing stable memory

Define:

$$\frac{dW_{\text{delay}}}{d\varepsilon} = \Phi(\varepsilon), \Phi(\varepsilon) > 0$$

$$W_{\text{delay}} = \int_{\varepsilon_*}^{\varepsilon} \Phi(u)\, du$$

$$W_{\text{commit}} = W_{\text{write}} + W_{\text{stabilise}}, W_{\text{write}} \geq k_B T \Delta I$$

Collapse Condition

$$\boxed{\int_{\varepsilon_*}^{\varepsilon} \Phi(u)\, du \;\geq\; W_{\text{commit}}(\varepsilon)}$$

Collapse occurs when maintaining uncertainty becomes more expensive than committing to structure, and at this point, ambiguity is no longer viable.

1.1 The DE Reality Equation

$$R(\varepsilon) = -\frac{dS}{d\varepsilon}$$

This condition can be expressed equivalently in terms of the entropy gradient, which defines the rate at which uncertainty can be converted into structure. This delay cost reflects the same gradient:

$$\Phi(\varepsilon) \sim -\frac{dS}{d\varepsilon}$$

Collapse therefore occurs when resolution becomes insufficient:

$$\boxed{R(\varepsilon) \leq R_c(\Upsilon)}$$

This is equivalent to the cost condition, where collapse is the loss of viable trajectories.

2. Outcomes of Collapse

Collapse always writes a local result. What differs is whether that result persists. Define:

$$B \in [0,1] \text{(alignment)}, R \in [0,1] \text{(retention)}$$

- B: how strongly memory selects a single basin
- R: whether the result survives recursion

Let $\{A_k\}$ be candidate attractors.

2.1 Annihilation

If retention fails:

$$R < R_{\min}$$

no memory is retained:

$$\Delta\Upsilon \approx 0$$

Collapse resolves locally but leaves no history.

2.2 False Attractor (Metastable)

If retention is sufficient but alignment is weak:

$$R \geq R_{\min}, B < B_{\min}$$

collapse produces a distributed state:

$$H(P) > 0$$

Local memory is written, but it is not coherently integrated. Under recursion, it cannot be retained:

$$\langle \Delta\Upsilon \rangle \approx 0$$

These states may persist locally, but they do not scale. Thus false attractors are structures that cannot be made self-consistent under recursive compression.

2.3 True Attractor

If both conditions are satisfied:

$$R \geq R_{\min}, B \geq B_{\min}$$

collapse concentrates into a single basin:

$$H(P) \approx 0$$

Memory is retained:

$$\Delta\Upsilon \gg 0$$

Structure propagates across scales. These outcomes correspond to annihilation, false attractors, and true attractors depending on retention and alignment.

$$\int \Phi(u)\, du \;\geq\; W_{\text{commit}} \Leftrightarrow R(\varepsilon) \leq R_c(\Upsilon)$$

Outcome:

- Annihilation → no retained memory
- False attractor → non-coherent memory (non-scaling)
- True attractor → coherent, persistent memory

4. Matter as Persistent Structure

A system exists only if it survives its own collapse. True attractors correspond to retained memory, and this persistence is what is observed as mass:

$$\text{mass} \propto \int (\text{retained structure})\, d\varepsilon$$

A particle is a stable solution of the collapse process, a self-sustaining attractor under recursive contraction.

5. Attractor Depth and Inertia

Let:

$$\Delta = D - \Lambda$$

where:

- D: basin depth
- Λ: disturbance load

Stability requires:

$$\Delta > 0$$

Mass reflects persistence across recursion. Inertia is the work required to reduce Δ. Deeper attractors resist change more strongly, whilst shallow attractors are easily perturbed.

6. Closure

The full chain is:

$$\text{collapse} \rightarrow \text{retained memory} \rightarrow \text{stable attractor} \rightarrow \text{matter}$$

Collapse is the point at which uncertainty ceases to be affordable. What follows depends on whether memory survives. To date memory has been treated as accumulated structure, the record of previous collapses. But not all memory is equally useful. What matters is whether that memory points clearly toward a single continuation, or whether it supports multiple competing possibilities. When a system approaches collapse, it must select a basin. If its stored memory strongly favours one option, the choice is clear. If the memory is ambiguous or internally inconsistent, the system has no single preferred direction and collapse distributes across alternatives.

We capture this distinction with the quantity B, which measures how decisively past structure selects a future state. A high value of B means memory is aligned and points to one basin. A low value means memory is spread across competing possibilities. To make this precise, we need a way to represent memory that is independent of how we describe the system. Memory must be defined in terms of what is preserved under recursion, not in terms of arbitrary coordinates. This leads naturally to a topological description.

Rather than treating memory as a list of stored states, we represent it as a set of persistent cycles, loops that cannot be removed by local deformation. Each loop carries a record of how the system has evolved. The effect of these loops can be measured through their holonomy, the net transformation accumulated when moving around a cycle. Some match the existing memory closely, others do not. The system therefore prefers attractors whose internal structure aligns with the holonomy of its stored memory. Bias B is then defined as the degree to which one attractor matches this memory better than all others. It is not an arbitrary parameter, but a measure of how strongly the past constrains the future.
In this way, alignment becomes a geometric property of memory itself, and collapse is guided by how well candidate structures fit the history already written.

Bias B as a holonomy-weighted memory alignment functional

Bias B is the degree to which stored memory singles out one attractor as the cheapest continuation. We represent memory by gauge-invariant holonomies around non-contractible cycles on the genus-g memory surface and measure, for each candidate attractor, how well its holonomy signature matches the current one. B is the gap therefore between the best and second-best matches, converted into a unit interval. Retention R is the probability that a committed state survives further recursion. We model this as a product of per-click survival probabilities across coarse-graining steps, where each step contracts entropic distance by α_{eff}. Collapse therefore has two gates, where B decides whether memory points to one basin, and R decides whether that basin can persist. Annihilation corresponds to low R, false attractor distributions correspond to low B with adequate R, and a true attractor corresponds to high $B$and high R.

Let the system's retained memory be represented by gauge-invariant loop content on the memory surface (your genus-g manifold). For a set of non-contractible cycles $\{\gamma_j\}_{j=1}^{2g}$, define the holonomy in the relevant internal sector:

$$U(\gamma_j) \;=\; \mathcal{P}\exp\left(\oint_{\gamma_j} A\right),$$

and compress it to a scalar "memory imprint" (any conjugation-invariant choice is acceptable; the trace is the simplest):

$$m_j \equiv 1 - \frac{1}{N}\Re(\mathrm{Tr}\, U(\gamma_j)) \quad \in [0,2],$$

where N is the representation dimension. $m_j = 0$ means "no retained twist" on that cycle, and larger m_j means stronger stored constraint. Now let $\{A_k\}$ be candidate attractor basins, each characterised by a preferred holonomy signature $m_j^{(k)}$ (the "template" that basin is compatible with). Define a mismatch energy in entropic space:

$$\mathcal{D}_k \equiv \sum_{j=1}^{2g} w_j \ (m_j - m_j^{(k)})^2, w_j \geq 0.$$

Then define bias B as how sharply memory discriminates between the best-fitting basin and the runner-up:

$$\boxed{B \equiv \sigma\left(\beta \ (\mathcal{D}_{(2)} - \mathcal{D}_{(1)})\right)}$$

where:

- $\mathcal{D}_{(1)}$ is the smallest mismatch across k,
- $\mathcal{D}_{(2)}$ is the second-smallest,
- $\sigma(x) = \frac{1}{1+e^{-x}}$ is a logistic squashing,
- β is an "inverse noise / salience" parameter.

If one basin matches memory much better than all others, $\mathcal{D}_{(2)} - \mathcal{D}_{(1)}$ is large, so $B \to 1$. If multiple basins match about equally well, the gap is small, so $B \to 0.5$ (or toward 0 if you shift the squashing). B is therefore a gauge-invariant, topology-aware measure of "how strongly the past points to one future." If we want $B \in [0,1]$with 0 meaning "no bias" for example, we use a centred version:

$$B \equiv \max\left(0,\ 2\sigma(\beta(\mathcal{D}_{(2)} - \mathcal{D}_{(1)})) - 1\right).$$

Retention R as the probability that commitment survives the ratchet, or whether the choice 'sticks' under recursion. Let our coarse-grained ratchet step be:

$$\varepsilon_{n+1} = \frac{\varepsilon_n}{\alpha_{\text{eff}}}, \Delta\Upsilon = \ln\ \alpha_{\text{eff}}.$$

A commitment must survive environmental noise, internal coupling stresses, and further coarse-graining. Define a stability margin for basin A_k:

$$\Delta_k \equiv D_k - \Lambda_k(\varepsilon),$$

where:

- D_k is basin depth (how strongly it holds structure),
- $\Lambda_k(\varepsilon)$ is the effective disturbance load at entropic depth ε.

Now define the per-click survival probability for the committed basin:

$$p_k(\varepsilon) \equiv \sigma\,(\gamma\,\Delta_k),$$

with γ setting sharpness. If commitment must survive n clicks to be macroscopic, define:

$$\boxed{R_k \equiv \prod_{m=0}^{n-1} p_k\,(\varepsilon_m)}$$

and take $R = R_{k\backslash *}$ for the basin actually selected. This definition uses the recursion index (clicks), not clock-time, so it makes retention a function of basin depth versus disturbance load. It can be made scale-aware because n can represent the number of coarse-graining layers required for the structure to become "classical."

A closed-form approximation tied to α_{eff}

If p_k is roughly constant across the relevant clicks (reasonable in a local regime), then

$$R \approx p^n.$$

Now connect n to the contraction itself. If each click removes a fraction $1 - \frac{1}{\alpha_{\text{eff}}}$ of accessible uncertainty, then the number of clicks required to compress uncertainty from εdown to $\varepsilon_{\backslash *}$is:

$$n \approx \frac{\ln(\varepsilon/\varepsilon_{\backslash *})}{\ln \alpha_{\text{eff}}}.$$

So:

$$\boxed{R \approx [\sigma(\gamma(D - \Lambda))]^{\ln(\varepsilon/\varepsilon_{\backslash *})/\ln\, \alpha_{\text{eff}}}}$$

Then larger α_{eff} means fewer clicks are required for compression, which increases R for a given per-click stability. But larger α_{eff} also means more aggressive pruning per click, which can increase Λ if pruning induces stress. So α_{eff} has a natural trade-off role. The parameter Λ represents the disturbance load placed on a committed structure during contraction. It measures the stress imposed by environmental noise and by the pruning process itself. Removing too many degrees of freedom at once generates structural strain, and Λ captures that strain. If the basin depth D exceeds Λ, the structure survives. If Λ exceeds D, the committed state destabilises. Thus α_{eff} plays a dual role as it accelerates resolution while simultaneously increasing the risk of structural fracture. Stable collapse therefore requires a balance between contraction efficiency and disturbance load. The disturbance term Λ in the DE collapse formalism plays an analogous structural role to the cosmological constant as both represent residual background stress opposing structural contraction. At cosmic scales, the surviving vacuum bias after entropic pruning may appear geometrically as the cosmological constant.

Chirality and Fractional Phase Structure

Chirality arises as a projection of topological phase-memory into geometric space. The *i-axis* carries phase and recursion, and when this structure is embedded into the geometric manifold, orientation becomes distinguishable. Left–right asymmetry therefore emerges from the ordering of phase under recursion. Spinors represent this structure algebraically, resolving orientation beyond simple phase. Their double-cover behaviour reflects the fact that geometric representation distinguishes orientation more finely than the underlying phase cycle. More generally, the recursion-phase state Υ evolves under a nonlinear map:

$$\Upsilon_{n+1} = f(\Upsilon_n)$$

governed by universal scaling constants. This produces stable fractional phase structures, not as new topological periods, but as attractors of the recursion. Examples include:

$$\frac{2\pi}{3}, \frac{\pi}{2}, \frac{3\pi}{2}$$

These values correspond to fixed points, bifurcation surfaces, or transition states in the recursion-phase geometry. They appear where the system approaches thresholds of stability, including chirality transitions, collapse surfaces, and entanglement boundaries. Fractional phase is therefore the signature of recursive structure approaching or residing on attractor manifolds in Υ-space. Fractional π attractors in Υ-space are central to the emergence of chirality, with the Feigenbaum constants governing the scaling structure of this map, as they do for all unimodal nonlinear recursions.

Quasi-attractors

Fractional π states arise as quasi-attractors of the recursion-phase map. These are intermediate states where the system approaches a fixed point without fully converging. The recursion operator:

$$R: (\varepsilon, \Upsilon, \theta) \mapsto (\varepsilon', \Upsilon', \theta')$$

generates scaling-dependent phase evolution, and near bifurcation thresholds this produces long-lived, non-closed cycles. These quasi-attractors are metastable structures in the evolution of Υ, and they correspond to fractional phase states such as:

$$\frac{\pi}{6}, \frac{\pi}{3}, \frac{5\pi}{6}$$

which act as intermediate stability points in the recursion. Structurally, quasi-attractors encode partial orientation and chirality, so generate metastable intermediate states that appear near collapse thresholds and bifurcation surfaces. They are the transitional geometry of the recursion, bridging reversible phase evolution and fully stable attractors. In this sense, quasi-attractors represent structure that is organised but not yet persistent, and they play a central role in the emergence of chirality and fine-scale structure within the DE framework.

Matter as Attractors in a Complex Information Field

In Penrose's formulation, matter arises as stable structures in twistor space. Spacetime is not fundamental but reconstructed from relations between complex objects. In

DE, this aligns naturally, where matter is not primary, but emerges as entropy-minimising attractors in a complex information field. Spinors provide the minimal representation of direction, phase, and helicity. Twistors extend this to relational structure, encoding how states connect without requiring spacetime as a starting point. Bi-twistors then represent separations between states, making them natural carriers of entropic distance ε.

In this sense, Penrose provides the mechanism by which structure forms in a complex field, while DE explains why such a representation is selected. A parsimonious system favours the smallest structure that maximises informational coherence, and spinors, twistors, and bi-twistors are precisely such structures. Matter therefore appears as stable attractors within this complex field. What we observe as particles and fields are persistent patterns that minimise informational cost while preserving coherence under recursion. This interpretation extends naturally to internal symmetries. Colour corresponds to memory constrained within closed loops, where internal states must sum to a neutral configuration. This produces confinement, because information circulates but cannot escape. Flavour, by contrast, represents memory across non-identical states. It encodes transitions rather than circulation, introducing ordering and irreversibility. Where colour stabilises structure, flavour allows it to evolve, thus colour is memory under closure and flavour is memory under transformation. The strong interaction preserves internal coherence, while the weak interaction updates the memory record. Together they define how structure persists and changes within the DE framework.

Two Universes, One Topology

The DE recursion admits two coherent orientation classes. A system with a U(1) phase fibre and a structural inversion rule naturally supports a pair of conjugate configurations, distinguished by the orientation of the phase-memory fibre. These configurations share the same geometric axes, the same topological rules, and the same scaling structure, they differ only in the direction in which recursion orders phase and selects attractors. In this sense, the laws of physics can be viewed as projections of a deeper topology, while chirality emerges as a global choice within that structure.

Spinors encode this distinction algebraically, and twistors express it geometrically. The asymmetry observed in our universe, including matter dominance, can therefore be understood as the selection of one orientation class over its conjugate. The alternative configuration remains a consistent, but unrealised, solution of the same underlying system. The topology does not favour one branch a priori, it admits both. What we observe is the consequence of which branch is selected under recursion.

In this framework, the wave function of the universe can be understood as a fundamentally dyadic structure embedded within this topology. The dyad encodes the space of possible orientations, while chirality determines how that space is traversed. A purely dyadic system preserves symmetry between alternatives. A chiral embedding breaks that symmetry just enough to select a direction. The universe we observe is therefore a resolved trajectory within a shared dyadic possibility space.

Chapter 5: The Higgs Boson

This image offers a visual interpretation of the Higgs boson, the particle associated with the field that gives mass to matter. The Higgs cannot be drawn in any literal sense, yet this illustration captures its conceptual structure through geometry and motion. At the centre sits a thick torus, a ring-like form that appears to ripple outward. This represents the Higgs field itself, a field that permeates all of space. The curvature and layered contours suggest a medium that is uniform yet internally dynamic. The Higgs field defines the background against which motion acquires cost.

Running through the torus is a vertical column, around which three intertwined bands spiral upward. These bands represent particle couplings to the Higgs field. Different particles interact with the field with different strengths, where stronger coupling corresponds to greater resistance to change (which we interpret as greater mass), and weaker coupling corresponds to lighter particles. In DE, this triadic structure is the minimal partition through which symmetry can break without losing coherence.

At the top and bottom of the column are opposing conical forms, which indicate symmetry breaking. Once the Higgs field settles into a stable, non-zero value, symmetry becomes unstable and the system selects a preferred orientation. From that point onward, motion acquires cost. The subtle chromatic offsets of cyan, magenta, and yellow introduce a sense of vibration. They suggest that the Higgs boson is not the field itself but a transient excitation within it. The boson appears as a ripple when the field is disturbed, briefly departing from equilibrium before decaying back into the vacuum state.

In the Deductive Entropy framework, the Higgs is the visible signature of a deeper informational transition. Mass is the entropic cost of preserving coherence once symmetry collapses. The toroidal centre represents a high-symmetry state where the entropy gradient is flat and all directions are equivalent. In that regime, no particle possesses mass because no direction is preferred. The vertical column represents the emergence of a gradient once symmetry destabilises, marking the transition from symmetric possibility to committed structure.

Mass arises when symmetry breaking constrains motion. In a fully symmetric system, degrees of freedom are unconstrained and motion carries no persistent cost. Once symmetry breaks, trajectories become restricted and motion requires energy to traverse an increasingly structured state space, and mass is the measure of that resistance. Within this frame, the Higgs field marks the transition at which excessive symmetry becomes unstable under recursion and must resolve into committed structure. The system lowers its uncertainty by selecting a structured configuration, breaking symmetry to preserve coherence. The Higgs field encodes this stabilisation, and the Higgs boson represents a local excitation of that field. Mass, therefore, is the energetic cost of maintaining structure after symmetry breaking, reflecting how strongly a particle is coupled to this stabilising field. In this sense, mass marks the transition from unconstrained recursion to persistent form, and it is this transition that allows the emergence of structure and hierarchy.

The discovery of the Higgs boson did not confirm the Standard Model in the simple way many anticipated. The model predicted the necessity of a scalar field to preserve renormalizability, allowing a broad range of values rather than precise mass. Naturalness arguments, radiative stability reasoning, and symmetry expectations led many physicists to anticipate a significantly heavier scalar than the one observed at approximately 125 GeV. In strict Popperian terms, detecting a scalar was a necessary condition for the theory, and in hindsight the precise value of its mass was the risky prediction. When the mass appeared well below many expectations, the result was absorbed as a discovery rather than the discrepancy interrogated. The narrative became "The Higgs is confirmed," rather than "The scalar exists, but our interpretation may be incomplete."

Three structural reasons explain this reaction. First, by the late 1990s the Higgs mechanism had become ontologically necessary within the Standard Model. Removing it would have destabilised large portions of quantum field theory, so that virtually any scalar that behaved roughly as expected would inevitably be identified as the Higgs. Second, naturalness quietly dissolved, whilst supersymmetry, compositeness and extra dimensions did not appear. Rather than revising foundational assumptions, the community largely retreated from naturalness itself. Third, large-scale experimental culture rewards detection more than falsification. The LHC was built to find something, and it was sufficient that a scalar was found.

Within this framework, however, the more important question is what role the scalar plays in the architecture of collapse. In DE the Higgs mass need not be treated as a free parameter, rather it is seen as a threshold cost tied to entropy reduction, memory locking, and scaling efficiency. Without memory there is no time, there is no before or after, and without memory, there is no way to distinguish one state from another as having occurred. We can have relations, correlations, configurations, but we cannot have order, and without order, words like before and after are meaningless. Before triadic lock-in, the system remains in a dyadic regime where correlations are reversible and no memory is retained. At the triadic threshold, correlation becomes unstable and commitment becomes entropically favoured.. That is where the Higgs mechanism operates, as an enforcement of commitment.

After lock-in engagement also becomes commitment, and relations are no longer optional as they carry consequences. Motion now resists change, and that resistance appears as mass. Chirality appears because the chosen configuration cannot be reversed without dismantling the structure that depends on it. Entanglement also changes character as it is no longer just a strong correlation that can be simply re-factored into independent components. It becomes a committed configuration (where separation carries real cost), which manifests as mass, inertia, decay pathways, and broken symmetries. Lock-in marks the transition from allowed correlations to enforced commitments.

With this transition history begins, parity can be violated, mirror states are excluded, and directionality emerges. Memory in this sense, is any retained asymmetry or constraint that persists through the transition. Once delay becomes more costly than selection, the system no longer explores the full space, rather it follows the path of least additional cost, which is the path already partially encoded. Collapse therefore becomes irreversible, and memory breaks time symmetry because it embeds prior constraint into the present state. Reversal would require erasing that structure, which is macroscopically prohibitive. The realised branch is the one most compatible with accumulated history, however slight that history initially was.

Across the Higgs threshold, this transition becomes enforced. Above it, electroweak degrees of freedom can defer commitment because delay is only weakly penalised. The symmetric state sits at an unstable maximum rather than a protected minimum, so small fluctuations are sufficient to push the system away from it. Once the threshold is crossed, descent becomes energetically favoured and delay becomes costly, so the vacuum selects a specific orientation. That selection reflects weakly encoded asymmetries arising from prior fluctuations.

The transition then is a change in curvature from a flat, degenerate surface to a sloped manifold where trajectories become history-dependent. Once mass appears, the resulting imbalance is expressed thermodynamically, but its origin is informational.

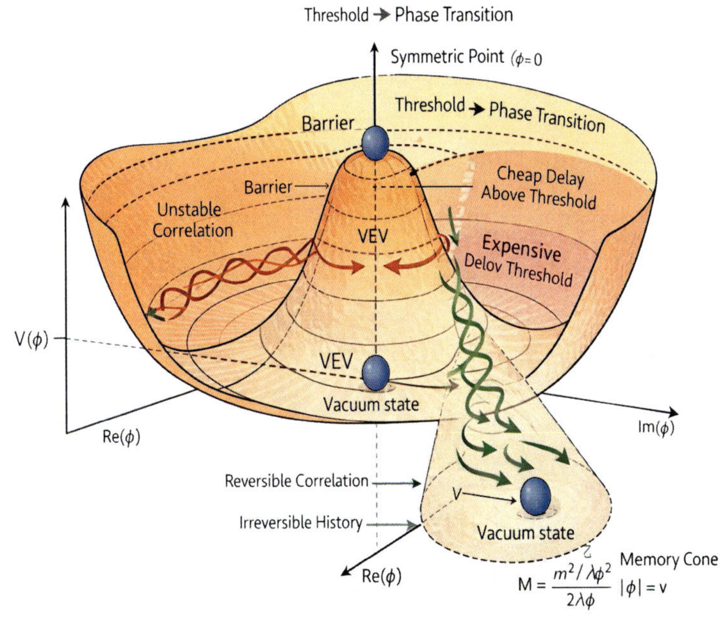

There is something unique about the Higgs that makes it a universal converter from correlation into memory. In conventional terms, the Higgs boson is unique because it has a vacuum expectation value, but ontologically it means the ground state itself carries information. The vacuum is no longer neutral and correlation can no longer float freely because the background has chosen. The Higgs potential is shallow, continuous, and the symmetric state is an unstable maximum, so it cannot retain the system once fluctuations push it away. Once the vacuum carries memory, every excitation must reference it. Mass, inertia, decay pathways, chirality, and irreversibility all inherit from that single act of commitment, no additional mechanism is required downstream. The Higgs then is the minimum structure capable of enforcing memory. It is the smallest possible intervention that converts reversible correlation into irreversible history. Anything weaker fails to lock in, while anything stronger collapses too early and destroys structure. The Higgs sits at the threshold where uncertainty can no longer remain uncommitted and must resolve into a persistent state. It occupies the minimal point at which enforcing commitment does not eliminate degrees of freedom faster than structure can form.

The Higgs is a scalar field whose lowest-energy state carries information. Most fields only store information when they are excited, so when they switch off and the effect disappears. The Higgs is different, because even when nothing is happening, its ground state is not neutral. It has a non-zero vacuum expectation value, meaning the vacuum itself is already committed. Hence it breaks symmetry without any force being applied, because the symmetric state is unstable. Remaining undecided costs energy and that converts correlation into an entropic liability, where delay becomes expensive. The Higgs couples universally, which makes its memory global rather than local, and once the vacuum remembers, everything else must remember too. As a scalar it has no direction, spin-0, and no orientation, and because it has no geometry of its own it can impose commitment without choosing a spatial direction.

Put together, the Higgs is the minimal memory-bearing substrate. It is the weakest possible field that still makes indecision unstable, and that is why it sits exactly at the threshold where correlation can no longer remain free. This also explains why its mass is small, and why its potential is shallow. It is tuned to the minimal condition under which the vacuum itself becomes a record. Once that happens, every level of reality inherits that commitment. Correlation becomes obligation, and possibility becomes memory.

So the Higgs boson does not explicitly choose matter over antimatter, it turns the vacuum into memory, and once memory exists, information is no longer symmetric. From that moment on, bias can accumulate. In a purely correlational universe, matter and antimatter are equivalent descriptions, and no outcome can be preferred because nothing is retained. Once the Higgs locks the vacuum into a committed state, information flow becomes path-dependent, and that is the precondition for asymmetry. For matter to dominate antimatter three conditions are required, known as the Sakharov conditions:

1. Baryon number violation
2. CP violation
3. Departure from thermal equilibrium

The Higgs transition provides the conditions under which these processes become effective. By breaking electroweak symmetry, it allows processes that violate particle number to proceed differently before and after collapse. By turning correlation into commitment, it makes CP-violating processes cumulative rather than self-cancelling, and by triggering a phase transition, it drives the universe out of equilibrium at exactly the moment when memory becomes unavoidable. Once memory exists, collapse direction matters, so that even a tiny asymmetry (statistical or quantum) is no longer averaged away. Instead it is preserved, amplified, and inherited. Matter survives because its informational lineage is reinforced by memory.

Antimatter does not disappear because it is forbidden, but because it fails to remain viable under committed history. The Higgs dyad collapses into a single memory-bearing branch. The conjugate branch still exists mathematically, but it no longer has entropic viability once memory is enforced. It cannot form an independent vacuum without undoing the very history that makes particles stable. This is why we do not observe an anti-Higgs vacuum competing with ours in this frame, because two simultaneous committed vacua would cancel memory and erase structure, so one must win in our branch. That also explains why CP violation becomes possible only after the Higgs threshold, once one branch is locked in.

Before the Higgs threshold, ε evolves symmetrically and ordering does not accumulate. After the threshold, ε becomes directional as retained constraint prevents reversal. Time therefore appears only once commitment is enforced. At the triadic threshold, correlation becomes unstable and commitment becomes entropically favoured. Matter and antimatter are symmetric at the level of correlation, where no history is retained and no outcome is preferred. Once memory becomes active, that symmetry is no longer neutral. Information is filtered through prior commitment, and one branch gains persistence because it is compatible with the accumulated constraint. The other becomes energetically disfavoured, and non-viable under committed evolution. The Higgs spatially activates this memory axis, which determines global polarity. Independent regions cannot sustain opposing memory structures without erasing the very constraints that define them. Competing histories cannot coexist stably, because memory is not local bookkeeping but a global condition on persistence. What the Higgs threshold marks is therefore the first moment at which memory becomes unavoidable as a property of the vacuum itself.

In this sense, the Higgs functions as a boundary condition. It is the minimal interface at which information ceases to be reversible and begins to be recorded. This parallels the role of a holographic boundary in gravitational systems, where information becomes encoded in a way that cannot be undone without loss. The Higgs is the local instantiation of that principle, the point at which correlation becomes record. Once this transition occurs, the physics of general relativity follows. Memory becomes primitive, mass enables persistence, and time emerges as the ordering of irreversible change. Gravity then appears as a secondary constraint, the geometric encoding of how committed information restricts future evolution. It is hence a bookkeeping rule, as once memory is written motion must respect it. This explains why gravity couples universally and why it is weak, because it reflects accumulated constraint rather than generating it.

Observationally this structure is consistent with the early universe. The Cosmic Microwave Background reveals a state that is smooth, isotropic, and low in gravitational entropy. Fluctuations are present, but they are not yet organised into persistent structure. This also aligns with Penrose's Weyl curvature hypothesis, where gravitational degrees of freedom are effectively inactive in the initial state. Following the Higgs transition, memory and mass exist, and ε acquires directional constraint, but gravity does not immediately dominate because structure is not yet localised. Correlation has collapsed into commitment globally, but not yet into distributed, persistent entities. Gravity emerges only when fermionic structures become ubiquitous, when memory is carried locally as well as globally. This expansion is from a phase transition across the entire system, a fermionic inflation where persistence is activated everywhere simultaneously. Once fermions carry memory locally, spacetime must respond locally. Geometry becomes dynamical, curvature becomes meaningful, and gravity emerges as the minimal encoding of distributed constraint. Early inflation smooths because correlation spreads before gravity can constrain it. Only after local commitment does curvature begin to encode information density.

The Memory Ratchet

The Higgs does not create memory ex nihilo, rather it ratchets memory that already exists at the smallest possible scale. If memory were absent before the Higgs, nothing could bias the collapse, no direction could be selected, and no branch could persist, so memory must be primitive. What the Higgs represents is the first macroscopic click of that primitive memory into durable history. At the Planck scale, there is already a minimal asymmetry, a seed memory that is not yet geometric, not yet temporal in the familiar sense, but nonetheless sufficient to break perfect reversibility. We can think of it as proto-memory, encoded as constraint rather than record, and the Higgs threshold is where that proto-memory becomes amplifiable, from a microscopic bias that cannot accumulate, where each fluctuation is almost perfectly erasable. It's at the Higgs scale that the system then gains that ability to 'lock in', and from that moment, every small asymmetry can be inherited. This is the ratchet that raises memory above the noise floor. Whilst the Planck scale provides the first bias, the Higgs provides the first irreversible click, and fermionic inflation provides the ubiquitous distribution of that click. Gravity appears only after enough of these committed states have accumulated to require geometric accounting. In that sense reality is constructed through successive ratchets of constraint. Linking the Higgs to

$$\Upsilon = i \, / \, \alpha$$

Feigenbaum's alpha is a canonical example of universal contraction near renormalisation thresholds. In it acts as an effective compression ratio governing when smooth variation gives way to discrete commitment. History grows only when primitive memory survives this compression, and the Higgs sits at the first scale where α becomes operative in a durable way. Below the Higgs scale, memory slips because correlation reabsorbs asymmetry, and the ϒ-value cannot grow. At the Higgs threshold, however, α clicks on, so degrees of freedom are forcibly reduced, the *i-axis* is no longer free to oscillate, and ϒ becomes non-zero for the first time. That is the first irreversible bifurcation of reality, and it's at this point where memory becomes cheaper than uncertainty, and the system transitions from reversible fluctuation to persistent structure.

Collapse, Vacuum Bias, and the Residual Curvature Term

If the Higgs threshold is the first scale at which commitment becomes cheaper than uncertainty, then the Higgs mass parameter encodes where that inequality stabilises. A dimensional estimate may be written schematically as

$$m_H \sim \frac{E_{\mathrm{EW}}}{\alpha c^2},$$

where E_{EW} is the characteristic electroweak energy scale and α is the effective entropic compression ratio governing durable pruning. This is a structural scaling claim rather than a precision derivation.

The Higgs, vacuum bias, and persistent energy

Once the Higgs field acquires a non-zero vacuum expectation value, the vacuum is no longer neutral. In conventional language this is symmetry breaking, more precisely the vacuum becomes biased because it acquires memory. Before this transition, the vacuum supports correlation without commitment where fluctuations exist, yet they are freely reversible. Possibilities arise and dissolve without consequence because nothing is retained. In such a regime, summing over all vacuum fluctuations as if they were persistent is physically misleading. Most of these states do not survive as history. The Higgs changes this by enforcing a threshold. Once the vacuum carries a preferred orientation, remaining undecided becomes entropically unstable. Some degrees of freedom are forced to commit, and that commitment can be retained as structure. From this point onward, the vacuum itself behaves as a record and it constrains which excitations can persist.

This reframes the vacuum energy problem. In standard quantum field theory, vacuum energy is often treated as a sum over zero-point contributions from all modes, leading to values vastly larger than the observed cosmological constant. In the DE framework, this overcounting arises because uncommitted possibilities are treated as if they were durable. Only modes that survive viability filtering (becoming part of the vacuum's retained memory) contribute to persistent structure and curvature. The cosmological constant therefore measures not the raw energy density of the vacuum, but the energy density of retained vacuum memory. It is non-zero because memory exists once the Higgs bias is established. It is extremely small because parsimony filters out almost all possibilities before they become history.

Dark matter fits naturally into the same picture, because if gravity is the geometric bookkeeping of committed information, then any structure that persists and resists change must gravitate, regardless of whether it couples electromagnetically. Dark matter may therefore correspond to memory-bearing structure that survives thresholding but does not participate in charge-based interactions. These three puzzles, vacuum energy, the cosmological constant, and dark matter, are usually treated as unrelated anomalies, but in DE they are different expressions of a single principle, namely that only committed structure persists.

Collapse as Entropic Viability

According to DE collapse is triggered when maintaining uncertainty becomes more expensive than committing to a stable record. That trigger is selected by entropic viability, where thermodynamic expense appears later as the macroscopic shadow of an informational inequality. Let ε denote entropic distance, a measure of unresolved uncertainty. Collapse is governed by a comparison of costs, maintaining unresolved uncertainty carries a cumulative delay cost, while committing to structure carries a one-off cost of writing and stabilising memory. The collapse condition is therefore:

$$W_{\text{delay}}(\varepsilon) \geq W_{\text{commit}}(\varepsilon)$$

Beyond this threshold, remaining undecided becomes more expensive than committing, and collapse is enforced. Two quantities determine the result:

• Bias B - whether memory strongly favours a single basin
• Retention R - whether the selected state survives further recursion

Bias reflects how well candidate states align with retained memory. If one basin matches significantly better than all others, collapse concentrates into that basin. If multiple basins are similarly compatible, collapse distributes across them. Retention reflects whether a committed state persists under recursive compression. Each step reduces accessible uncertainty by a factor α_{eff}, while survival depends on the balance between basin depth $D$and disturbance load Λ. To leading order:

$$R \approx [\sigma(\gamma(D - \Lambda))]^n$$

where n is the number of effective compression steps.

Residual vacuum stress and the cosmological constant

The disturbance term $\Lambda_k(\varepsilon)$ introduced above is local and dynamical. It is not identical to the cosmological constant of general relativity, and the similarity is structural rather than literal. In Einstein's field equations,

$$G_{\mu\nu} + \Lambda_{\text{GR}}\, g_{\mu\nu} = 8\pi G\, T_{\mu\nu},$$

Λ_{GR} acts as a uniform background curvature term, behaving like a persistent vacuum pressure. It is global and approximately constant. In DE, most vacuum fluctuations remain uncommitted possibilities, as they do not survive viability filtering and therefore do not contribute to persistent structure. Only a small residual bias becomes durable after the Higgs locks the vacuum into a committed orientation. That surviving bias is weak and globally distributed because it is embedded in the vacuum itself. This motivates the mapping:

$$\Lambda_{GR} \propto \langle \Lambda_{DE} \rangle_{cosmic},$$

where Λ_{DE} denotes the residual, scale-averaged disturbance load associated with vacuum commitment after entropic pruning. The cosmological constant is therefore not the sum of all vacuum modes, it is the residual of what persists after uncertainty filtering. In this view, Λ_{GR} is the curvature signature of incomplete vacuum commitment, the small irreducible stress that remains once the universe has learned to remember.

Chapter 6: Deriving the Born Rule

Consider a system described not by a single outcome, but by a waveform of possibilities. Each component of the wave carries both magnitude and phase, and it is the phase that determines how different components combine. When two components are brought together they interfere of course. Alignment produces reinforcement, misalignment produces cancellation, and the resulting structure depends on how these phases combine across the whole.

At the smallest scales, physical systems behave in exactly this way. What exists prior to observation is not a definite state, but a superposition of admissible configurations, each contributing to a shared wave of possibility. Measurement does not reveal a pre-existing value. It selects a configuration that can persist once these possibilities are forced to resolve. The rule that connects this wave description to observation is simple and precise. If a system is described by an amplitude ψ, the probability of observing a given outcome is proportional to

$$| \psi |^2$$

This prescription predicts interference patterns, scattering outcomes, and statistical distributions with remarkable accuracy, yet its origin is not obvious. Why should probability depend on the square of the amplitude rather than the amplitude itself? Why should phase combine linearly, but observable outcomes depend quadratically? These questions reflect a deeper constraint on how structure can persist when multiple possibilities overlap and must eventually resolve into a single, stable configuration.

For nearly a century, this has remained an assumption at the heart of quantum mechanics. It is one of the most successful rules in science, and one of the least explained. In this chapter, we take a different approach, by asking what must happen for any structure to persist in a world built from overlapping possibilities? And we will see that probability is indeed the visible trace of some deeper principle, that relates directly to the survival of coherent structure in a field of competing possibilities.

Whilst quantum mechanics remains one of the most successful theories ever constructed, one of its central components remains unexplained. The Born rule states that the probability of observing a system in a given state is:

$$P_i = |\psi_i|^2$$

where;

P_i: probability of outcome i

ψ_i: complex probability amplitude for outcome i

$|\psi_i|$: modulus (magnitude) of the complex amplitude.

This rule seems to work perfectly, as it predicts interference, measurement outcomes, and statistical distributions with extraordinary precision. But it is not derived from deeper principles. It is introduced as a postulate, which creates a fundamental conceptual gap. The Schrödinger equation governs the continuous evolution of the wavefunction, yet the transition from amplitude to probability is imposed externally. Why should probability be proportional to the square of amplitude? And why not linear dependence, or some other function? Here in the DE framework, probability is not taken as fundamental, rather it is seen as a direct consequence of recursive stability. What appears as probabilistic behaviour on the surface is simply a visual projection of a deeper, hidden selection process.

In the DE formulation, phase is the primary variable and structure arises through the accumulation and stabilisation of phase. Let the system be described by an entropic field:

$$\Psi(\chi, \varepsilon, i)$$

where:

- χ: configurational coordinate
- ε: recursion depth (ordering coordinate)
- i: memory axis (phase coherence dimension)

The observable component of the system is the subset from the field that survives recursive stabilisation. This is governed by the operator Υ, which selects phase-consistent structures. Thus, the relevant quantity is the coherence of phase across recursion. Consider a decomposition into candidate states:

$$\Psi = \sum_i \psi_i$$

Each component ψ_i represents a possible configuration. However, not all configurations persist, as stability requires that a configuration remains coherent across recursive cycles. Define a coherence measure along the memory axis:

$$\mathcal{C}_i = \int_{\mathcal{V}_i} \rho_i \,(\chi, \varepsilon)\, d\mu$$

where:

- ρ_i: phase coherence density
- $\mathcal{V}_i$: region of phase space supporting state i
- $d\mu$: measure over entropic space

Only regions that maintain coherence under recursion contribute to observable structure.

The Quadratic Stability Constraint

Recursive systems impose a key constraint, namely that stability must be preserved under superposition and recombination, and this requirement is highly restrictive. So consider two overlapping states $\psi_1$and ψ_2. The total field is:

$$\Psi = \psi_1 + \psi_2$$

The total coherence must include both individual contributions and their cross-term interference. Thus:

$$\mathcal{C}_{\text{total}} \propto | \psi_1 + \psi_2 |^2$$

This is a limiting consequence of phase addition, interference structure and recursive preservation of coherence. Linear measures fail because they do not conserve coherence under interference. Only the squared modulus preserves additivity and positivity, whilst preserving stability across recursion. Thus, the quadratic form emerges as the unique stable measure of recursive phase coherence. We can now define probability as the fraction of total coherent phase volume associated with each state:

$$P_i = \frac{\mathcal{C}_i}{\sum_j \mathcal{C}_j}$$

Using the quadratic stability constraint:

$$\mathcal{C}_i \propto |\,\psi_i\,|^2$$

we obtain:

$$P_i = \frac{|\,\psi_i\,|^2}{\sum_j |\,\psi_j\,|^2}$$

which is precisely the Born rule. In this framework the wavefunction represents a field of unresolved possibilities, and the *i-axis* encodes phase memory and coherence. The operator Υ selects stable recursive structures, so probability is the measure of how much of the phase space associated with a configuration remains coherent under recursion, and states with greater coherent support are more likely to persist through collapse.

Collapse as Threshold Selection

Collapse occurs when the system can no longer sustain coherence across multiple configurations:

$$S(\Psi) > R(\varepsilon)$$

At this threshold, only configurations satisfying recursive stability survive, and the probability of each outcome reflects its coherent phase volume prior to collapse. Thus, collapse is seen to be deterministic filtering under recursive constraints, with probability emerging from the geometry of the coherence space. The square arises because phase combines linearly, coherence must remain positive, and interference must be preserved. As stability must survive recursive composition, the conditions uniquely select:

$$|\psi|^2$$

as the only valid measure. Therefore in this frame quantum probability is the visible consequence of this deeper process, where phase is primary and structure arises from stabilised phase. As stability requires coherence across recursion, coherence is preserved only under a quadratic measure. Thus:

$$P = |\psi|^2$$

is a necessary consequence of recursive phase stability. If coherence is to be preserved, only one measure survives recursive combination:

$$|\psi|^2$$

and so the Born rule is selected by an emergent spacetime as the minimal condition under which possibility can become persistence. If every possible outcome exists as a pattern in phase, and a candidate for existence, it is only those patterns that remain coherent across recursion can endure, and the remainder dissolve back into raw possibility. The square function measures that endurance, it tells us how much of a configuration can carry itself forward into memory, and hence into what we finally observe as structure.

Quantum mechanics therefore describes a world in which structure is rare, coherence is fragile, and only the most stable patterns survive. In this frame probability is the shadow of that process, it is what remains when the universe tests all possibilities and keeps only those that are recursively viable.

Uniqueness and Physical Interpretation of the Quadratic Measure

At this point, a natural question arises. We have argued that probability reflects the coherent phase support of a configuration under recursive evolution, and that this leads to a quadratic dependence on amplitude, but is this form uniquely required, or merely convenient? To answer this, consider what any physically meaningful measure of observable outcomes must satisfy. Let $M(\psi)$ be the measure associated with a candidate state ψ. For this measure to correspond to observable probability, it must obey four minimal constraints:

1. **Positivity**

$$M(\psi) \geq 0$$

   since probabilities cannot be negative.

2. **Normalization**

$$\sum_i M(\psi_i) = 1$$

   over mutually exclusive outcomes.

3. **Stability under Superposition**
   If two configurations combine $\psi = \psi_1 + \psi_2$, the measure must preserve interference structure rather than destroy it.

4. **Additivity for Orthogonal States**
   For mutually exclusive (non-interfering) configurations,

$$M(\psi_1 + \psi_2) = M(\psi_1) + M(\psi_2)$$

when cross-terms vanish. Now consider the most general combination:

$$\psi = \psi_1 + \psi_2$$

Any admissible measure must produce:

$$M(\psi) = M(\psi_1) + M(\psi_2) + \text{interference terms}$$

The interference terms are essential, because without them, the theory cannot reproduce experimentally observed phenomena such as the double-slit pattern. The only class of measures that preserves interference, remains positive, and is stable under repeated composition, is a quadratic form:

$$M(\psi) = \langle \psi, \psi \rangle = | \psi |^2$$

Explicitly:

$$| \psi_1 + \psi_2 |^2 = | \psi_1 |^2 + | \psi_2 |^2 + 2\text{Re}\,(\psi_1^* \psi_2)$$

This expression preserves cross-term interference, reduces correctly when states are orthogonal, and remains positive.

Excluding Alternative Measures

Suppose now instead we attempted:

$$M(\psi) \propto | \psi |^p$$

for some $p \neq 2$.

Then:

- For $p = 1$:
  interference structure is lost or misrepresented. The measure becomes linear and cannot reproduce observed interference fringes.

- For $p > 2$:
  normalization becomes unstable under superposition, and compositional closure fails. Small

amplitudes are suppressed too strongly, distorting statistical outcomes.

- For non-polynomial forms either positivity or additivity breaks, or the measure fails to remain stable under recursive recombination.

Thus, the quadratic form is the unique measure that preserves coherence under superposition while remaining stable under recursion. To complete the argument, we must show that the quadratic form is not merely sufficient, but necessary. Let $M(\psi)$ be any candidate measure assigning observable weight to a state ψ. For M to correspond to physically meaningful outcomes, it must satisfy four constraints:

1. Positivity

$$M(\psi) \geq 0$$

2. Normalisation

$$\sum_i M(\psi_i) = 1$$

3. Superposition Consistency

For $\psi = \psi_1 + \psi_2$ the measure must preserve interference:

$$M(\psi) = M(\psi_1) + M(\psi_2) + \text{interference terms}$$

4. Recursive Stability

The measure must remain invariant under recursive composition:

$$M(R(\psi)) \propto M(\psi)$$

where R is the DE recursion operator.

Exclusion of Alternatives

Assume a general form:

$$M(\psi) \propto |\psi|^p$$

Case 1: $p = 1$

$$M(\psi_1 + \psi_2) = |\psi_1 + \psi_2|$$

Cross-terms are not preserved correctly, and interference patterns are distorted, which violates empirical observation

Case 2: $p \neq 2$

For $p > 2$ small amplitudes suppressed too strongly, so measurement becomes unstable under recombination

For $p < 2$ cross-terms dominate, positivity and normalization fail under recursion

Case 3: Non-polynomial forms

Any nonlinear functional $M(\psi) = f(|\psi|)$ fails at least one of:

- additivity for orthogonal states
- interference preservation
- recursive closure

Quadratic Uniqueness

Only the quadratic form $M(\psi) = |\psi|^2$ satisfies all these conditions because:

$$|\psi_1 + \psi_2|^2 = |\psi_1|^2 + |\psi_2|^2 + 2\mathrm{Re}\,(\psi_1^*\psi_2)$$

which is the unique bilinear form compatible with linear superposition. In simple terms amplitude encodes phase coherence, recursion filters unstable configurations, where only stable coherence regions persist. Thus $|\psi|^2$ measures surviving coherence under recursion, not simply abstract probability.

Coherence as a Measure of Persistence

Within the DE framework, this quadratic form has a direct interpretation. Each configuration occupies a region in recursive phase space. The amplitude ψ_i encodes the local phase coherence of that region. The squared modulus:

$$|\psi_i|^2$$

measures the extent of coherent support that survives recursive evolution. In other words the square quantifies how much of a configuration can remain stable across recursive composition. A further question then arises, that why should this coherence measure correspond to observed frequencies in repeated experiments? The answer follows from the structure of recursive sampling. Each experimental trial prepares the system under the same constraints, probing the same entropic phase structure. Over many trials, the system repeatedly resolves into stable configurations drawn from the same coherence landscape. Thus, the frequency with which outcome i occurs approaches:

$$P_i = \frac{\mathcal{C}_i}{\sum_j \mathcal{C}_j}$$

where $C_i \propto |\psi_i|^2$ is the coherent phase measure of that configuration. This is directly analogous to classical statistical mechanics, where phase-space volume determines probability. Here, the relevant space is not classical configuration space, but recursive phase coherence space. In standard quantum mechanics, the Born rule is introduced as an axiom:

$$P_i = |\psi_i|^2$$

Here in the DE framework, it is derived as a consequence of three facts:

1. Phase is the primary variable
2. Structure arises from recursive stabilisation of phase
3. Stability requires preservation of coherence under superposition

Thus the Born rule is selected by the conditions required for structure to persist. Probability is the measure of recursive phase coherence that survives collapse, and the square appears because only quadratic coherence is stable under the recursive composition of possibilities. Thus the Born rule tells us how much of any specified outcome can endure, and emerges as the sampling measure of recursively stable phase regions. Thus the novelty here is that the observable weighting of outcomes is determined by recursive stability rather than postulated as an axiom.

Chapter 7: Memory Fields

Standard physics treats fields as primitive, with particles arising as their excitations. In the DE framework, this order is reversed. Memory is fundamental, understood as the persistence of constraint across recursion, and fields arise as the minimal structure required to make that memory consistent across space. If a system carries memory as an internal record $\psi(x)$, that record may be represented differently at each point without changing its physical content. This local freedom implies that neighbouring records cannot be directly compared. To make comparison meaningful, the system must specify how a record at one point is transported into the local frame of another before any difference is taken. This requirement forces the introduction of a connection $A_\mu(x)$, which defines how internal memory is transported while preserving identity.

Once transport is defined, differences between neighbouring records become well-posed. However, transporting a record around a closed path need not return it to its original form. The residual misalignment accumulated around such loops is curvature. The stationarity condition that minimises this misalignment yields Maxwell and Yang–Mills equations as the lowest-cost configuration of memory transport. Fields are therefore not primitive objects, but the minimal bookkeeping required for consistent memory under recursion.

1) Primitive: memory as a constraint over histories

In DE, the fundamental object is what remains consistent under recursion, rather than "state at t". Let a system carry a local "internal record" $\psi(x)$ (phase, orientation, multi-component order parameter). Memory penalises inconsistency between neighbouring records, and this penalty is quantified as the entropy of unresolved mismatch. So we introduce a memory-tension functional over space:

$$\mathcal{M}[\psi] \;=\; \int d^4x \;\| \nabla\psi(x) \|^2$$

This would suffice if records were globally comparable, but DE allows local representational freedom.

2) Gauge freedom is forced by memory

Memory is relational and if the "record" has an internal phase or orientation, then the same physical memory can be represented in many locally equivalent ways:

$$\psi(x) \;\rightarrow\; U(x)\,\psi(x)$$

where $U(x)$ is a local symmetry transformation (U(1), SU(2), SU(3)). Then $\nabla\psi$ loses meaning, because it compares records in mismatched local frames. This forces the introduction of a connection:

$$D_\mu\psi \;=\; (\partial_\mu + iA_\mu)\psi(\mathrm{U}(1))$$

or in general

$$D_\mu\psi \;=\; (\partial_\mu + A_\mu)\psi, A_\mu \in \mathfrak{g}$$

Memory constrains local freedom (we can store the same relational record in different gauges), whilst also demanding global consistency (we must be able to transport a record without contradiction). Therefore the minimal device that satisfies both is a field A_μ, where fields appear as the bookkeeping required for consistent memory transport.

3) Curvature measures loop inconsistency

Once you allow local frame changes, moving a record around a loop can fail to return it unchanged. That failure is curvature:

$$F_{\mu\nu} = \partial_\mu A_\nu - \partial_\nu A_\mu (\mathrm{U(1)})$$

or

$$F_{\mu\nu} = \partial_\mu A_\nu - \partial_\nu A_\mu + [A_\mu, A_\nu] (\text{non-abelian})$$

Where;

- $F_{\mu\nu}$ is loop-inconsistency of memory.
- A "field strength" is literally a measure of irreducible memory mismatch that cannot be gauged away.

4) The DE action: minimal cost of memory consistency

The system penalises two quantities:

1. mismatch in the memory record under transport, $| D\psi |^2$
2. mismatch generated by the connection itself, $| F |^2$

The minimal scalar functional is:

$$S[\psi, A] = \int d^4x \, (\alpha \, (D_\mu \psi)^\dagger (D^\mu \psi) + \beta \, \mathrm{Tr}(F_{\mu\nu} F^{\mu\nu}) + V(\psi))$$

This has the standard form, but its origin differs. In the conventional view, symmetry implies a gauge field. In DE, comparability of memory under local freedom forces the connection, and the action is the minimal cost required to maintain consistency.

5) Varying the memory action produces field equations

Vary S with respect to A_μ leads to:

U(1) case (Maxwell form)

$$\partial_\nu F^{\nu\mu} = j^\mu$$

Where $$j^\mu \propto \mathrm{Im}(\psi^\dagger D^\mu \psi)$$

So Maxwell's equations emerge as the stationarity condition of memory-consistency, where F evolves to reduce loop-inconsistency, and sources j^μ are places where the memory record ψ forces nonzero transport structure.

Non-abelian case (Yang–Mills form)

$$D_\nu F^{\nu\mu} = J^\mu$$

These equations describe the evolution of the connection toward configurations that minimise loop inconsistency. Sources arise where the memory record ψ enforces non-trivial transport structure. Maxwell and Yang-Mills equations are therefore Euler-Lagrange equations of minimising memory inconsistency.

6) Gravity in the same DE template

The same logic applies when the "record" is local spacetime orientation. Memory of geometric alignment requires a connection (the spin connection), and loop inconsistency becomes Riemann curvature. Minimising curvature cost yields Einstein-type equations in the coarse-grained limit, and gravity is therefore the memory field of geometry.

Fields are the minimal compensators required for a universe that permits local representational freedom while preserving coherent memory under recursion. Gauge curvature is the irreducible residue of memory mismatch, and Maxwell and Yang–Mills equations are the stationarity conditions of minimising that mismatch. In this sense, gauge sectors are memory-stabilising structures. Bosonic fields define tolerance envelopes for persistent configurations, whilst the Higgs sets the boundary for stable chirality. Configurations outside that boundary decohere because they are entropically inefficient.

From memory transport to physical fields

If a system preserves memory under local representational freedom, then a connection is forced, and curvature emerges as irreducible mismatch. The resulting stationarity conditions yield the familiar field equations. What remains is to identify how these abstract memory constraints manifest in the physical world. In the Standard Model, this structure appears as the gauge sectors and their associated fields. From the DE perspective, these are specific implementations of memory stabilisation under different constraints. Each gauge sector can therefore be interpreted as enforcing coherence within a particular domain of internal degrees of freedom. The question is no longer why these fields exist, but what aspect of memory they stabilise.

1. Nuclear fields as memory fields

Where SU(3) confines colour, SU(2) organises chirality and weak transitions. In DE language, we can reimagine that SU(3) stabilises relational orientation in colour space under extreme compression, whilst SU(2) stabilises handedness under recursion in phase space. In both cases, the gauge structure prevents arbitrary drift because it enforces allowable transformations and excludes others. In that sense a memory field is one that constrains how internal degrees of freedom may change without destroying persistent structure. Confinement in QCD is particularly interesting here as colour cannot exist freely. Any attempt to separate it increases energy nonlinearly, which looks exactly like a memory-tension penalty for breaking relational coherence. So the strong interaction behaves exactly like a high-density memory stabiliser.

2. Bosons as tolerance operators

In this frame gauge bosons then become the mediators that enforce tolerance limits. In conventional theory gluons enforce colour charge conservation, whilst the W and Z bosons enforce weak isospin transitions. But now we can reimagine that bosons define the allowable error bandwidth of the memory sector. They transport constraint between locations, restoring coherence when local deviations occur. Hence gauge bosons become regulators of admissible variation within a memory sector.

3. The Higgs as a chirality boundary

In the Standard Model the Higgs is a scalar field. Its vacuum expectation value breaks $\mathrm{SU}(2)_L \times \mathrm{U}(1)_Y$. It gives mass to fermions via Yukawa coupling, and it distinguishes left and right components. So the Higgs defines a boundary between massless chiral propagation and massive coupled propagation. Before symmetry breaking fermions are purely chiral with no rest mass, after left and right components are locked and mass emerges. That locking can be reframed, as the Higgs sets the recursion cost at which chirality can remain decoupled. Above that threshold left/right remain separate and propagation is lightlike, but below that threshold, left/right lock into a massive attractor and time-like propagation emerges.

4. Decoherence as competitive inefficiency

Anything outside the Higgs-defined tolerance decoheres and is outcompeted by more efficient solutions. In the DE frame "non-optimal" states are outcompeted in a Darwinian sense, yet a configuration persists only if the total memory-cost functional is stationary and bounded. If a state increases curvature cost excessively, it breaks gauge consistency and violates chirality coupling conditions, and hence it decays. Configurations exceeding tolerance thresholds are dynamically unstable and flow toward lower-cost attractors.

Therefore the nuclear gauge sectors clearly function as memory stabilisation fields (even if not recognised conventionally). Their bosons define the tolerance envelopes within which internal degrees of freedom may vary while preserving coherent recursion. The Higgs field establishes the chirality boundary at which left and right components become energetically locked. Configurations that exceed these tolerance limits are dynamically unstable and relax toward lower-cost attractor states. The deep implications of this are that the strong force is high-density memory confinement, whilst the weak force is chirality bookkeeping. Higgs sets the recursion-to-mass transition threshold and electromagnetism is long-range phase memory. Gravity then becomes the geometric memory of stress-energy.

Fields as Elastic Substrates

Fields arise because preserving distinguishability under recursion with local representational freedom forces an alignment connection, penalising loop inconsistency forces curvature cost. This is the stationary point of "distinguishability loss + loop frustration" yields Maxwell/Yang–Mills, and the Higgs vacuum sets the chirality-locking tolerance that selects which persistent fermionic attractors can exist. Field theory struggles most where interactions are strongly coupled, history-dependent, or non-perturbative. It is exactly where memory, constraint accumulation, and structural inertia dominate. In those regimes, the idea that a field is a smooth bookkeeping device for instantaneous forces starts to look like the wrong abstraction.

Quantum field theory predicts scattering amplitudes extremely well when disturbances are small and local. It degrades when disturbances reshape the background itself. That is what you expect if a field stores strain rather than transmits force, whereas elastic media respond linearly only near equilibrium. Renormalisation then makes sense if short-scale disturbances permanently deform the medium and must be absorbed into effective parameters. If fields were forces, divergences would be pathological. If fields are elastic substrates, renormalisation is simply coarse-graining stored stress. Entanglement and vacuum correlations look odd if influences must propagate force-like through spacetime. They look natural if the field already contains pre-strained structure, where disturbance reveals correlation. In condensed matter, elasticity descriptions outperform microscopic force models for most situations. The same pattern appears in physics more broadly, but effective field theories work because they describe how the medium responds. From that perspective, fields are closer to constitutive relations than to forces, they encode how reality yields under constraint.

This also explains why alternative frameworks often outperform field theory in specific domains. Lattice methods, holographic dualities, agent-based models, and information-theoretic approaches all implicitly treat the substrate as something with memory and stiffness, when they stop pretending the vacuum is inert. Field theory works best when the buffer is barely deformed, and it fails when deformation becomes dominant. In this frame fields then become emergent stress maps of a deeper constraint-accumulating substrate, not fundamental forces acting in empty space.

Here is a compact set of concrete examples where the elastic buffer interpretation fits better than force-based fields.

• **Quantum Chromodynamics**
Perturbative QCD works at high energies and fails at low energies. Confinement and mass gaps emerge only through lattice methods. That behaviour matches an elastic medium that stiffens as strain accumulates, not a force that simply weakens or strengthens.

• **Renormalisation group flow**
Couplings change with scale because short-range disturbances permanently deform the background. This looks like stress being absorbed into material properties, not a fundamental force 'changing its mind'.

• **AdS/CFT correspondence**
Bulk gravity dynamics map to boundary field deformations. The bulk behaves like a strained geometric medium whose response encodes boundary information. Forces are secondary, because stored structure does the work.

• **Superconductivity and phonons**
Electrons pair through lattice deformation rather than direct

attraction. The lattice acts as an elastic buffer that reshapes interaction pathways. This is a clear physical precedent for fields as response media.

• **Vacuum polarisation in QED**
The vacuum screens charge and alters propagation constants. That is textbook elastic behaviour. A force picture has to bolt this on after the fact.

• **Effective field theories in general**
Low-energy theories outperform high-energy "fundamental" ones because they describe how the substrate responds once it has already absorbed historical constraint.

Across these cases, prediction improves when fields are treated as maps of stored deformation rather than as primitive forces. That pattern is consistent, repeatable, and hard to unsee once noticed.

Appendix A: Nuclear Fields as Memory fields

1) DE primitives

- **Memory record**: a local internal state $\psi(x)$ that must remain comparable under recursion.
- **Local freedom:** $\psi(x) \to U(x)\psi(x)$ where $U(x)$ is a local change of representation.
- **Memory transport**: comparisons must be made with a connection A_μ so the comparison is frame-consistent. The DE "memory cost" starts as:

$$S_{\mathrm{mem}}[\psi, A] = \int d^4x \ ((D_\mu\psi)^\dagger(D^\mu\psi) \ + \ \lambda \, \mathrm{Tr}(F_{\mu\nu}F^{\mu\nu}))$$

with

$$D_\mu = \partial_\mu + A_\mu, F_{\mu\nu} = \partial_\mu A_\nu - \partial_\nu A_\mu + [A_\mu, A_\nu]$$

Where;

- $D_\mu \psi$: local memory mismatch (how hard it is to keep the record coherent across space).
- $F_{\mu\nu}$: loop inconsistency (irreducible "memory frustration" around closed transport).

This is already "nuclear fields = memory fields" in a precise way, with gauge curvature is the measurable residue of mismatch that cannot be gauged away.

2) Higgs as the chirality tolerance boundary

Starting from the Standard Model chiral split, with a fermion with left and right components:

- ψ_L transforms as an $\mathrm{SU}(2)_L$ doublet.
- ψ_R is an $\mathrm{SU}(2)_L$ singlet.

A **direct mass term** $m\bar{\psi}\psi = m(\bar{\psi}_L\psi_R + \bar{\psi}_R\psi_L)$ is not gauge invariant, so chirality cannot "lock" without an intermediary that carries the right transformation property. That intermediary is the Higgs doublet H. The Higgs therefore is the $\mathrm{SU}(2)_L$that allows left and right "records" to become comparable without breaking $\mathrm{SU}(2)_L$ consistency. In action form, chirality-locking is a memory coupling term:

$$S_{\text{lock}} = \int d^4x \, (y\, \bar{\psi}_L H \psi_R + y^* \, \bar{\psi}_R H^\dagger \psi_L)$$

This is the only gauge-consistent way to "suture" the left and right memory streams.

The tolerance boundary is the vacuum expectation value

Let the Higgs potential be

$$V(H) = -\mu^2 H^\dagger H + \lambda (H^\dagger H)^2$$

Minimising gives

$$\langle H^\dagger H \rangle = \frac{\mu^2}{2\lambda}, v \equiv \sqrt{2\langle H^\dagger H \rangle} = \frac{\mu}{\sqrt{\lambda}}$$

Then the effective fermion mass is

$$m_f = \frac{y\, v}{\sqrt{2}}$$

Where;

- v is the critical tolerance at which chirality can remain decoupled (massless propagation) or becomes energetically locked (massive propagation).
- The Yukawa coupling y is a locking efficiency parameter, or how strongly a given fermion species commits to that lock. So the "Higgs at chirality limit" becomes a concrete inequality about stable propagation regimes:
- Chiral (effectively unlocked) when $E \gg yv$
- Locked (massive attractor) when $E \lesssim yv$

3) Decoherence

Define a DE functional that measures the total cost of maintaining coherent recursion for the fermion sector:

$$\mathcal{C} = \int d^4x \, [\underbrace{(D_\mu \psi_L)^\dagger (D^\mu \psi_L) + (D_\mu \psi_R)^\dagger (D^\mu \psi_R)}_{\text{memory transport cost}} + \underbrace{\kappa \mid y\bar{\psi}_L H \psi_R \mid}_{\text{locking cost / benefit}} + \underbrace{V(H)}_{\text{tolerance landscape}} + \underbrace{\lambda_g \mathrm{Tr}(F^2)}_{\text{curvature frustration}}]$$

Stable matter configurations are stationary points of $\mathcal{C}$, whilst states "outside tolerance" are those for which the coupled Euler–Lagrange equations have no bounded stationary solution, so they decay (decohere) toward an available attractor. If a configuration is not an extremum (or is a saddle with decay channels), it relaxes. Lower cost attractors dominate the path integral weight, whilst higher memory-cost histories are exponentially suppressed;

$$\text{weight} \sim e^{-S_{\text{eff}}}$$

4) Why bosons are "tolerance operators"

The gauge bosons are excitations of the connection A_μ, and their role is to enforce local comparability of ψ under allowable transformations. So A_μ is the memory-transport operator, and $F_{\mu\nu}$is the memory-frustration density. The bosons set the "allowed bandwidth" because they determine how quickly deviations are corrected and what deviations are permitted (via the group structure).

Fields are the minimal bookkeeping required for a universe that preserves coherent memory under local representational freedom. Gauge bosons are tolerance operators, excitations of the connection that enforce admissible variation in internal degrees of freedom. Curvature is measurable memory frustration, the residual mismatch around closed transport. The Higgs field defines the chirality tolerance boundary and without it, left and right chiral records cannot be coupled without violating $\mathrm{SU}(2)_L$ consistency. When the Higgs acquires a vacuum expectation value v, chirality becomes energetically lockable with strength $m_f \propto yv$. Histories that exceed the tolerance envelope are unstable as they fail to form bounded stationary points of the memory-cost functional and relax toward available lower-cost attractors.

Appendix B: Higgs as Chirality Locking Operator

1) Start with what DE really means by memory

Let $\psi(x)$ be the local "record" (multi-component, complex, chiral). In DE, memory is the capacity to carry distinguishable structure forward through recursion. So the primitive object is a recursion operator R_ε, mapping the current representation to the next effective description:

$$\psi \;\mapsto\; \psi_\varepsilon \equiv R_\varepsilon[\psi]$$

If recursion were perfect, distinguishability would be preserved, but in reality, coarse-graining, coupling, noise, and competing constraints degrade it.

2) Define a memory-cost as distinguishability loss

Choosing a scalar that measures "how much structure got lost" between ψ and its recursively updated form. A clean choice is a KL divergence between local state distributions, or equivalently Fisher information in the small-step limit. Represent the local record as a family of probability densities $p(z \mid x)$ parameterised by $\psi(x)$ (this is standard in information geometry, where ψ indexes the distribution). Then define:

$$\mathcal{C}_{\text{mem}} \;=\; \int d^4x \; D_{\text{KL}}\left(p(z \mid x, \psi) \;\|\; p(z \mid x, \psi_\epsilon)\right)$$

For small recursion steps, KL expands to a quadratic form in parameter shifts:

$$D_{\mathrm{KL}} \approx \frac{1}{2}\, \delta\theta^T\, \mathbf{I}(\theta)\, \delta\theta$$

where $\mathbf{I}$ is the Fisher information metric and $\delta\theta$ is the effective parameter change induced by recursion. In this frame Fisher information is the local "distinguishability stiffness", and memory cost is the penalty for blurring distinguishability under recursion.

3) Local freedom forces a connection

But the record is only defined up to local reparameterization (gauge freedom). That means:

$$\psi(x) \to U(x)\psi(x)$$

If we try to compare $\psi(x)$ at neighbouring points, we must specify how we identify frames, otherwise the "difference" is meaningless. So we introduce an alignment field $A_\mu(x)$ that performs the minimal frame reconciliation between x and $x + dx$. Defining the recursion-consistent transport:

$$\psi(x + dx) \;\approx\; \exp\left(-A_\mu dx^\mu\right) \psi(x)$$

Then the true mismatch is not $\partial_\mu \psi$, it is the mismatch after best alignment, i.e. a covariant derivative:

$$D_\mu \psi \equiv \partial_\mu \psi + A_\mu \psi$$

This step is forced by the requirement that memory be comparable under local representational freedom.

4) Why the gauge kinetic term appears

If we can choose A_μ freely at every point, we could always set mismatch to zero by overfitting the alignment. That would "save memory" locally while creating global inconsistency. DE calls that a false attractor, local coherence purchased by global contradiction. The diagnostic of global contradiction is loop failure, so transport around a closed loop must return the record consistently. The obstruction is curvature:

$$F_{\mu\nu} = \partial_\mu A_\nu - \partial_\nu A_\mu + [A_\mu, A_\nu]$$

The DE memory functional must penalise two things:

1. **local distinguishability loss** (mismatch after alignment)
2. **global overfitting** (curvature, the residue of inconsistent alignments)

This yields the DE action in its natural form:

$$S_{\mathrm{DE}}[\psi, A] = \int d^4x \, [\, \underbrace{(D_\mu\psi)^\dagger \, \mathbf{G}(D^\mu\psi)}_{\text{distinguishability loss term}} + \underbrace{\lambda \, \mathrm{Tr}(F_{\mu\nu}F^{\mu\nu})}_{\text{anti-overfit / loop-consistency term}} + V(\psi)]$$

G is the information-geometry metric (Fisher metric) induced by your chosen family $p(z \mid x, \psi)$.

5) Euler–Lagrange gives Maxwell/Yang–Mills as a memory minimum

Varying A_μ yields:

$$D_\nu F^{\nu\mu} = J^\mu$$

where the current J^μ comes from the $(D\psi)^2$ term and is the flow of distinguishability constraint. In U(1) it reduces to Maxwell form:

$$\partial_\nu F^{\nu\mu} = j^\mu$$

Field equations are then the stationarity conditions of least distinguishability loss + least loop frustration.

6) The Higgs as the tolerance boundary

Now split the record into chiral streams:

- ψ_L lives in a different representation ($\mathrm{SU}(2)_L$ doublet) than ψ_R(singlet).
- They cannot be compared directly without breaking the memory-consistency rules.

So DE demands a bridge record H with the right transformation properties to reconcile them. The unique renormalisable bridge is the Yukawa lock:

$$S_{\text{lock}} = \int d^4x \ (y \, \bar{\psi}_L H \psi_R + \text{h.c.})$$

Now give H a tolerance landscape:

$$V(H) = -\mu^2 H^\dagger H + \lambda (H^\dagger H)^2$$

The minimum defines the tolerance radius:

$$v = \frac{\mu}{\sqrt{\lambda}}$$

Chirality locking strength becomes:

$$m_f = \frac{yv}{\sqrt{2}}$$

So v is the chirality coherence tolerance of the vacuum and y is the locking efficiency. When the recursion energy scale

E falls below yv, the cheapest way to preserve memory is to lock L and R into a massive attractor. Collapse outcomes fall into three distinct classes:

(i) Annihilation (collapse fails to propagate)

No bounded stationary point exists for S_{DE} given constraints, so recursion cannot carry distinguishable structure forward. The record decays, no persistent excitation.

(ii) False attractor (local coherence, global inconsistency)

You can reduce $(D\psi)^2$ locally by choosing A_μ to overfit, but curvature explodes and F^2 grows. These are metastable, noisy, decohering configurations.

(iii) True attractor (stable matter/field configuration)

A joint minimum exists where both mismatch and curvature are bounded. That solution defines a stable field sector with bosons as tolerance excitations, fermions as memory-bearing excitations, Higgs as the chirality boundary condition.

Chapter 8: Renormalisation

The previous chapter explored the idea that fields are structured ways of preserving constraint across scales. Spin preserves constraint locally, and electromagnetism transports it without entropy cost. Time then emerges as the ε-weighted accumulation of committed history. The next question then becomes unavoidable, because if memory accumulates, why does it not diverge?

Renormalisation is traditionally described as a method for relating physical descriptions across scales. In the Deductive Entropy framework, it acquires a deeper and more physical meaning. It is the process by which systems convert transient correlations into persistent structure. Each renormalisation step integrates out microscopic degrees of freedom and replaces them with effective variables that encode prior interactions. What survives coarse-graining is accumulated memory. The fundamental variable is the ordering coordinate of emergence ε, which parameterises progression through the recursion. The stabilised memory of the system is then defined as

$$\Upsilon(\varepsilon) = \int_0^{\varepsilon} \varepsilon' \, d\varepsilon'$$

From this perspective, renormalisation is the mechanism by which accumulated memory is created. At each scale transition, correlations that survive stabilisation are encoded into Υ, increasing the depth of memory. The system thereby retains a compressed record of its prior states. A molecule carries the memory of atomic constraints, a cell carries the memory of chemistry, and each effective theory carries the memory of the scale beneath it.

This interpretation introduces a natural distinction between attractors. Not all attractors support continued recursion. A viable or "true" attractor is one that preserves sufficient accumulated memory to allow further emergence. It stabilises correlations without eliminating the degrees of freedom required for subsequent bifurcations. In contrast, a "false" attractor either over-constrains the system or fails to stabilise it. In the first instance, accumulated memory becomes rigid and cannot support further organisation. In the second, insufficient memory is retained and no stable structure forms. In both cases, recursion cannot proceed.

The defining property of a true attractor is therefore its ability to scaffold further accumulation of memory. By scaffolding, we mean the capacity to preserve and transmit structured correlations such that additional levels of organisation can form. Each successful renormalisation step must achieve a balance, by compressing uncertainty into stable memory while preserving enough freedom to seed the next bifurcation. This balance is the condition for scalable complexity. Renormalisation thus marks the boundary at which possibility becomes commitment. It is the point at which the system selects a direction along ε and encodes that selection into Υ. What survives becomes the substrate for further recursion, and what is discarded is removed from the accessible state space.

In this formulation, complexity emerges as the cumulative effect of successive memory-forming transitions. Each level of organisation exists because the previous level preserved just enough accumulated memory to support it. The universe therefore builds upon what it retains. This leads directly to the dynamical picture developed in this chapter. If renormalisation is the process that creates accumulated memory, then the evolution of that memory must follow a coherent law. In the DE framework, that law takes the form of an oscillator in accumulated memory,

$$\frac{d^2\Upsilon}{d\varepsilon^2} = -\Omega^2\Upsilon$$

with ε defining the orientation of motion and the stabilisation operator Υ governing the persistence of its cycles. The universe can then be understood as a recursive system in which ε drives emergence, Υ accumulates memory, and structure arises from the continuous interplay between the two. Thus the universe is not created at a singular moment, it is continuously generated through the unfolding of ε. The apparent origin corresponds to the observer's perspective of crossing the $\varepsilon = 0$ surface, while cosmic expansion reflects the ongoing separation of the two chiral branches of the same underlying process.

In the idealised limit this appears harmonic, but under recursive constraint and memory accumulation, the oscillator becomes progressively damped and biased, producing directional evolution rather than closed cycles;

$$\frac{d^2\Upsilon}{d\varepsilon^2} + \gamma\frac{d\Upsilon}{d\varepsilon} + \Omega^2\Upsilon = 0$$

Where:

- $\gamma > 0$ encodes memory / irreversibility

When Collapse Becomes Cheaper Than Chaos

Every time a system commits structure, it sacrifices degrees of freedom. That sacrifice is the price of persistence, so for instance a molecule admits fewer admissible configurations than its constituent atoms, and a crystal fewer again. Even a biological cell has far fewer degrees of freedom than the chemistry beneath it, and at each stage, freedom is reduced so that identity can survive. Renormalisation is the disciplined re-articulation of a physical theory that accompanies the selective elimination of microscopic degrees of freedom when stable macroscopic structure becomes energetically preferable. In conventional quantum field theory, renormalisation appears as a technical adjustment. Divergences arise at small scales, infinities threaten predictions, and parameters must be rescaled to recover finite quantities, so hence it can look procedural. In the DE frame, it occurs physically whenever macrostate collapse becomes energetically cheaper than continued microstate entropy growth. When the cost of tracking microscopic freedom exceeds the cost of committing higher-order structure, the system reorganises. Irrelevant degrees of freedom are integrated out, and a new effective description emerges. That transition is renormalisation.

The Scaling Constraint

Scaling is governed by parsimony. In earlier chapters we established that uncertainty follows the entropic geodesic in state space, so recursion advances along trajectories that minimise entropy expenditure per unit of committed structure. What appear superficially as "constraints" on scaling are in fact the geometric consequences of that minimisation. Consider an increment of committed memory Δi, producing

$$\Delta\Upsilon = \frac{\Delta i}{\alpha}$$

When a system explores possibilities, its state is described by information and for scaling to persist, this increment must lie on a viable geodesic in uncertainty space. Many configurations are still possible, so the system carries uncertainty, but once one configuration stabilises and persists through recursion, that information is no longer just a possibility, it becomes part of the structure of the system. In the equations this is represented as a unit of committed constraint. There are several common situations where this conversion appears;

First is **memory formation.**
A system begins with uncertain states that can be described informationally. When a pattern becomes reliable and survives noise, it becomes memory. At that point the information is no longer fluctuating, it has become a fixed constraint within the system's structure. In the equations this looks like information collapsing into a stable unit.

Second is **symmetry breaking.**
In many physical systems there are initially many equivalent possibilities. When symmetry breaks, one configuration is chosen and the others disappear. The choice becomes embedded in the geometry of the system and what was once informational uncertainty becomes a structural feature of space. For example, once a crystal lattice forms, the positions of atoms are no longer informational variables. They are part of the structure.

Third is **renormalisation in scale transitions.**
When a system moves from one scale to the next, many

microscopic degrees of freedom are compressed into a single macroscopic parameter. Large numbers of informational states are replaced by a smaller set of effective variables. In other words, a whole region of informational detail becomes a single structural constraint at the higher scale.

Fourth is **learning in biological systems**.
Before learning, many neural firing patterns are possible. After learning, synaptic weights stabilise certain patterns, where the information that once described possibilities is now encoded as physical connectivity. Again, informational variability becomes structural commitment.

The general principle is that information describes possibilities, while structure describes commitments. Whenever a system stabilises a possibility so that it persists across time and recursion, the informational description collapses into a structural one. In symbolic terms this often looks like information being converted into a unit of committed structure. In the DE framework this conversion is what allows scaling to proceed, as each step of recursion takes uncertain informational microstates and turns some fraction of them into stable constraints that define the next level of organisation.

1. Memory commitment

Information becomes stable constraint.

$$\Delta\Upsilon = \frac{\Delta i}{\alpha}$$

Meaning a small amount of information that stabilises produces a proportional increase in structural commitment.

2. Structural compression across scales

Degrees of freedom shrink when the system renormalises.

$$\Upsilon = \frac{1}{\alpha}$$

So only a fraction of the previous possibilities survives the transition to the next level of structure.

3. Entropic resolution

$$H_{\varepsilon+1} = H_{\varepsilon} - i$$

where

H = entropy
i = resolved information

Where each step along the ε ordering removes uncertainty as information stabilises.

4. Structural accumulation

$$S_{\varepsilon+1} = S_{\varepsilon} + 1$$

So each successful commitment adds one unit of constraint to the system's structure.

5. Renormalisation of degrees of freedom

$$N_{\varepsilon+1} = \frac{N_{\varepsilon}}{\alpha}$$

Many microscopic possibilities compress into fewer effective structural variables.

6. Entropic viability condition

$$\frac{dH}{di} < 1$$

So structure forms only when committing information reduces entropy faster than maintaining uncertainty. Scaling occurs because uncertainty resolves.

7. Attractor locking

Once information stabilises, it defines the attractor of the system.

$$A_{\varepsilon+1} = f(A_\varepsilon + 1)$$

So that each commitment reshapes the landscape in which future commitments occur, and stabilised information then becomes the geometry.

Feigenbaum's α

In the dyadic phase, uncertainty is preserved and degrees of freedom remain open. The system delays commitment and this preserved uncertainty carries selective advantage because it maintains flexibility. Multiple futures remain admissible, correlations can adjust, so decisions are deferred. In the triadic collapse, uncertainty resolves into structure, where degrees of freedom are sacrificed, and a memory is written. Constraint appears and the system becomes committed. Scaling depends on the tension between these two regimes. If uncertainty is preserved too aggressively, structure never stabilises. If commitment is imposed too early, flexibility disappears and recursion halts. In nonlinear dynamics, at the accumulation point of period-doubling bifurcations, the ratio between successive scale intervals converges to Feigenbaum's constant

$$\alpha \approx 2.5029.$$

This value sits precisely at the boundary between smooth variation and discrete hierarchy, marking the precise scaling ratio at which dyadic preservation and triadic commitment remain in balance.. If α were smaller, too little freedom would be sacrificed per scale, so that excess degrees of freedom would remain active, causing memory to leak back into correlation, and hierarchy would fail to stabilise. If α were larger, too much freedom would be destroyed, where structure locks in prematurely, flexibility vanishes, and recursion freezes. Thus α represents the minimal entropy tax that still guarantees inheritance across scales. It is the most efficient local scaling rule compatible with unbounded recursion.

At α each ratchet click removes just enough uncertainty to stabilise structure while preserving enough residual freedom to seed the next bifurcation. It is the minimal entropy tax that guarantees inheritance across scales, so it is the most efficient local scaling rule compatible with unbounded recursion. Viable hierarchy emerges at the boundary where uncertainty is preserved long enough to remain adaptive, yet sacrificed soon enough to become memory.

Dimensional Collapse and the Metastable-Attractor Transition

Thus the transition from metastable oscillation to configuration lock can be understood as a dimensional collapse in the regulatory state space. Before the transition, biomarker dynamics occupy a relatively high-dimensional region in which multiple regulatory modes fluctuate semi-independently. During the collapse transition, cross-marker correlations strengthen and system motion becomes dominated by a smaller number of coupled modes. In linear algebra terms, the covariance matrix develops a dominant eigenvalue and effective dimensionality contracts. Let the biomarker state vector be

$$x(\varepsilon) = (x_1, x_2, \ldots, x_n)$$

with covariance matrix

$$\Sigma_{ij}(\varepsilon) = \mathrm{Cov}(x_i, x_j).$$

Define effective dimensionality using the standard participation-ratio estimator:

$$d_{\mathrm{eff}}(\varepsilon) = \frac{(\sum_{i=1}^{n} \lambda_i (\varepsilon))^2}{\sum_{i=1}^{n} \lambda_i^2 (\varepsilon)},$$

where λ_i are the eigenvalues of the covariance matrix. The dyad corresponds to relatively high d_{eff}, meaning variance is distributed across many independent modes. Collapse corresponds to a contraction

$$d_{\mathrm{after}} = (1 - \eta) d_{\mathrm{before}}$$

where η is the fractional loss of independent degrees of freedom. The trajectory architecture implies that this collapse occurs during the transition between the metastable oscillatory phase and the synchronized acceleration phase. So the effective reduction in degrees of freedom at renormalisation is 40%. Let us now consider alternative values for renormalisation.

Scenario A: 35% Dimensional Contraction

If

$$\eta = 0.35,$$

then

$$d_{\mathrm{after}} = 0.65 d_{\mathrm{before}}.$$

The system retains a larger fraction of its regulatory freedom, so independent subsystems remain partially decoupled and oscillatory exploration continues for longer. In phase-space terms the instability manifold is broad, where trajectories wander within a metastable basin, and correlations increase more slowly. Entropy rises gradually and configuration lock is delayed because coupling between axes remains incomplete. This corresponds to a system that searches extensively for compensatory configurations before committing to collapse.

Scenario B: 45% Dimensional Contraction

If

$$\eta = 0.45,$$

then

$$d_{\text{after}} = 0.55 d_{\text{before}}.$$

So the system loses almost half of its independent regulatory modes during the transition, correlations strengthen rapidly and independent oscillations collapse into a small number of coordinated modes. In phase-space terms the instability manifold narrows, so the oscillatory exploration is brief, and trajectories converge quickly toward the attractor. This corresponds to a system that reorganizes abruptly once compensatory regulation fails.

Regulatory architectures with this intermediate contraction regime are advantageous because they provide two competing properties simultaneously:

1. **Adaptive search capacity.**
   The system retains sufficient degrees of freedom

during stress to explore compensatory configurations.

2. **Rapid reorganization.**
   When collapse becomes unavoidable, coupling between subsystems is strong enough to produce rapid synchronization and transition into a new attractor.

Architectures with too little contraction remain overly diffuse and unstable, whereas architectures with too much contraction lose flexibility prematurely. Systems in the intermediate regime maintain both adaptability and decisive reorganization. Thus the observed contraction range is consistent with the structure of evolved regulatory networks, where the contraction of approximately 35–45% in effective dimensionality provides the dynamical behavior

If we model the transition with a surviving complexity fraction of 0.35, then the system retains a large exploratory search space during the metastable phase and then undergoes a severe dimensional collapse at lock-in. In DE terms this produces a long oscillatory phase, high entropy production, and a sharp final synchronization into the attractor state. Evolutionarily, this architecture makes sense because it preserves adaptive search under stress while still allowing rapid coordinated reorganization once continuation in the old state is no longer viable. The practical consequence is that the most informative signal lies in trajectory instability during the metastable window, not in absolute marker abnormality.

$$d_{\text{after}} = 0.35\, d_{\text{before}}$$

implies

- 65% loss of independent regulatory variance
- metastable oscillatory exploration before transition
- maximal entropy production in the pre-collapse window
- sharp renormalization into a low-dimensional attractor

If we now imagine that the contraction value were 0.35 instead of 0.40, which at first glance the difference seems tiny, in reality it would have enormous consequences because the contraction happens again and again at every level of organisation, and these small differences compound across recursion. The most immediate consequence would be that systems would compress information more aggressively. Each new level of structure would eliminate more possibilities from the previous level. The system would stabilise patterns quickly, but it would also lose flexibility more rapidly, so that innovation would become constrained. When too many possibilities disappear at each step, systems struggle to explore alternative configurations, and consequently the evolutionary search becomes narrowed. Biological evolution, technological innovation, and cultural change would all move more slowly because fewer viable paths would remain open.

Second, systems would lock into patterns earlier. With stronger contraction, stable structures form quickly, but they also become rigid, and once a pattern appears, it is harder to escape from it. Complex systems would therefore fall into attractors more readily.

Third, large-scale complexity would be rarer. The growth of complexity depends on preserving enough freedom at each level so that the next level can form. If too much freedom is removed too early, the recursive ladder stops. Complex hierarchies such as living organisms, ecosystems, or intelligent brain would struggle to emerge.

Fourth, systems would be more stable but less adaptable. Structures would resist disturbance because they would be tightly constrained. However, they would also adapt poorly to changing environments. Once conditions shifted, many systems would simply fail rather than reorganise.

Finally, the universe would appear simpler. There would likely be fewer levels of nested organisation. Biological diversity would be lower, cognitive systems would be less flexible, and the overall richness of emergent structures would be reduced.

In short, a contraction value of 0.35 would favour order over exploration. Systems would stabilise quickly but would lose the freedom needed to generate deep layers of complexity. The universe would still form structure, but it would produce fewer of the intricate, adaptive systems we associate with life and intelligence. Ultimately such a regime would be evolutionarily unstable relative to more balanced architectures (viz. 0.4 d_{after}).

Now imagine the opposite direction, where the contraction value is 0.45. Again at first this sounds like a small change, but because scaling happens repeatedly across many levels as mentioned previously, the effects would accumulate throughout complex systems. At 0.45, the system shifts further toward coordination and stability, away from deep exploratory search.

$$d_{\text{after}} = 0.45\, d_{\text{before}}$$

The transition causes a 55% contraction in effective dimensionality, less severe than 0.35 and slightly less than 0.40. That means more independent variance survives the transition, and compared with 0.35 and 0.40. Correlations would strengthen earlier, but not as brutally as the system retains more flexibility after transition. The metastable oscillatory phase is shorter and less dramatic, thus the final attractor is less tightly locked. So the system tends to explore less deeply before committing, reorganize earlier, and preserve more residual adaptability after reorganization. The most important difference is that more possibilities would survive each step of organisation. Systems would retain greater freedom as they move from one scale to the next, and this would produce several consequences.

First, systems would explore more possibilities before stabilising because less information is compressed at each step, and patterns would take longer to become fixed. Evolution, learning, and adaptation would involve a broader search through potential configurations.

Second, structures would be more flexible but less stable because more degrees of freedom survive, the resulting patterns would be easier to modify. Systems would adapt readily to new conditions, but they would also be more prone to fluctuation and instability.

Third, recursion could extend to deeper levels of complexity, thereby preserving more freedom means that higher levels of organisation have more raw material to build upon. Complex hierarchies might become richer, with more branching pathways and more diversity of outcomes. However, this advantage comes with a trade-off, because if too many possibilities remain open, systems struggle to commit to stable structure. Noise and fluctuation can overwhelm emerging patterns before they have time to stabilise. Instead of forming coherent attractors, systems may drift through many partial configurations. In such a universe, complexity might be abundant but less coherent. Structures could be creative and adaptable, but also fragile and short-lived. Biological systems might display enormous diversity but less long-term stability, and cognitive systems might be highly flexible but prone to instability. Large-scale organisation might exist, yet remain constantly fluid.

In simple terms, a contraction value of 0.45 would favour exploration over stability making structure harder to stabilise, whereas 0.35 compresses too strongly, leading to rigid systems. The value around 0.40 d_{after} appears to sit in a delicate middle ground, because it removes enough freedom to stabilise structure, yet preserves enough possibility for new levels of organisation to emerge. That balance may indeed be one of the quiet conditions that makes a complex universe possible.

Living systems are organized close to critical transitions, where they can remain stable but are capable of rapid reconfiguration. Most of the time biological systems operate in high-dimensional regulatory space, but during stress they enter metastable oscillatory regimes and when instability exceeds buffering capacity they collapse into lower-dimensional attractors. This is exactly the pattern seen in immune activation, metabolic collapse, ecological regime shifts, neurological transitions, and even carcinogenesis. If this scaling reflects evolved network architecture, then many biological transitions should show the same structure;

1. stable regime
2. oscillatory metastability

3. rapid collapse into a new attractor

In this interpretation, diseases such as cancer arise when the same adaptive architecture that protects organisms under stress reaches its limits. The system behaves exactly as designed, exploring compensatory states, redistributing regulatory load, and delaying transition. Once buffering capacity is exhausted, it reorganizes into a new attractor, which may be pathological.

Living systems move continuously along a complexity axis. Growth and development increase regulatory dimensionality, while stress and disease reduce it, with recovery restoring lost structure. Life is therefore a dynamic cycle between high-dimensional exploration and lower-dimensional coordination. This pattern is consistent across developmental biology, neural plasticity, immune learning, and ecosystem resilience.

As a system approaches collapse, the metastable phase corresponds to maximal entropy production. Metabolic cost rises, signalling turnover increases, and multiple regulatory subsystems compete simultaneously. Clinically, this produces the characteristic confusion of pre-collapse states, where no single dominant pattern has yet emerged.

Evolution favours regulatory architectures that balance these competing demands. Hierarchical feedback, modular coupling, and nonlinear control produce systems that can explore widely under stress yet reorganize rapidly when necessary. Such systems operate near the edge of instability, preserving complexity while retaining the capacity for decisive transition. Life persists by repeatedly negotiating this boundary between exploration and collapse.

Before transition, variance is distributed across many weakly coupled modes, and effective dimensionality is high. After collapse, variance concentrates into a small number of dominant modes, correlations lock, and dimensionality contracts sharply. The critical question is how much contraction occurs.

A viable adaptive system must satisfy two opposing constraints, retaining sufficient degrees of freedom to explore alternative configurations, whilst it must also be capable of rapid synchronization once transition becomes unavoidable. Empirically, this balance occurs when approximately 60% of variance becomes coupled, leaving roughly 40% independent:

$$d_{\text{after}} \approx 0.4\, d_{\text{before}}$$

Nonlinear dynamical systems approaching bifurcation exhibit universal geometric scaling governed by the Feigenbaum constant:

$$\alpha \approx 2.5029, \frac{1}{\alpha} \approx 0.399$$

This ratio appears across systems undergoing period-doubling cascades, independent of their physical substrate. Its proximity to the observed contraction fraction suggests that the balance between exploration and coordinated collapse is not incidental, but a universal property of systems evolving under recursive constraint. The ~0.4 fraction therefore emerges naturally as the condition under which adaptive complexity can both persist and reorganize.

Renormalisation Reinterpreted

Local scaling efficiency does not imply unbounded growth. In the DE framework, scaling is constrained globally by the cosmological constant Λ. A positive Λ defines a de Sitter horizon with finite entropy, imposing a maximum achievable history depth:

$$\Upsilon \leq \Upsilon_{\max}(\Lambda)$$

This introduces a dual constraint on scaling. The local contraction ratio α governs viability at each step, while Λ sets the global boundary on accumulated memory. Together, they ensure that complexity can grow without divergence. Renormalisation occurs at the threshold where maintaining microscopic freedom becomes more expensive than committing to structure. For a system with microstate entropy S_{micro} and macrostate commitment cost C_{macro}, this condition is:

$$\frac{dS_{\text{micro}}}{d\varepsilon} > \frac{dC_{\text{macro}}}{d\varepsilon}$$

At this point, the system integrates out high-frequency degrees of freedom, shifts to an effective description, and encodes the surviving correlations as stable structure. This transition is renormalisation. A viable attractor is one that supports continued scaling under the global constraint:

$$\Upsilon < \Upsilon_{\max}(\Lambda)$$

while preserving transmissibility of structure across scales. If the system over-collapses, it loses the degrees of freedom required for further organisation. If it under-collapses, it fails to stabilise. In both cases, recursion cannot proceed. Renormalisation therefore marks the entropic boundary at which collapse becomes cheaper than continued microscopic resolution, while remaining consistent with the global limit set by Λ.

The diagram illustrates this transition, where successive compression steps converge toward invariant structure until either viable scaling or collapse is reached."

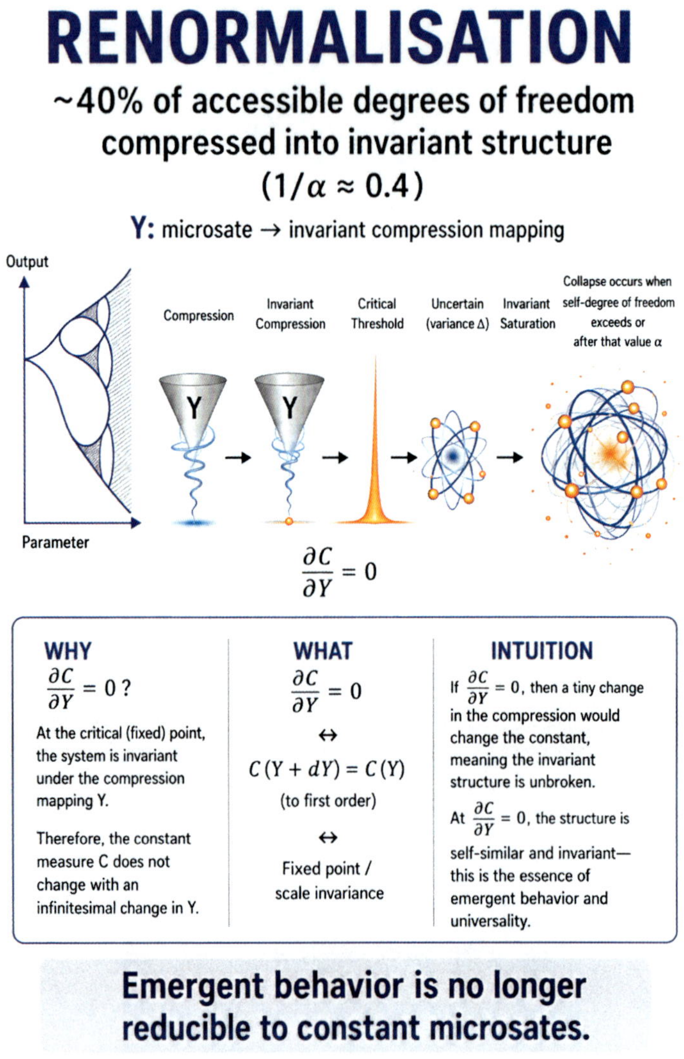

Time as the Conjugate of Entropy Polarity

Hence renormalisation shows that scaling proceeds by selective sacrifice of degrees of freedom, and at each threshold microscopic detail is integrated out so that stable macroscopic structure is retained. What then survives becomes history. Scaling therefore produces accumulated commitment, such that each renormalisation step writes structure into reality. If scaling is the geometry of viable hierarchy, then time is the ledger of its cost. Every coarse-graining, every ratchet click, every reduction of freedom carries an energetic price, and that price is what converts transient correlation into durable memory.

Previously we have shown time to be geometrically emergent from the primitive entropic distance ε as a topological axis. Now we must consider the effect of entanglement and complexity on that emergence. Let Υ denote committed history depth, let ε denote the marginal energetic cost of extending that history.

$$\tau \equiv \int \varepsilon \, d\Upsilon.$$

Time emerges when structure stabilises across scale as accumulated history weighted by its energetic cost, where Υ measures how much history has been committed, and ε measures the expense of each extra unit of history. Time is thus the integral of ε over Υ, meaning time is generated exactly when history grows, weighted by the energetic price of that growth. "Time dilation" arises because different regions pay different ε for the same increase in history, whilst the cosmological constant sets a hard ceiling on how much history Υ can ever accumulate, which in turn sets a ceiling on how much time can ever be produced. This definition carries immediate consequences. If no history accumulates, $d\Upsilon = 0$, then no time is produced, and if no recursion budget exists, $\varepsilon = 0$, again no time is generated.

Time is therefore computed from state variables. Using the earlier identity

$$\Upsilon = \frac{i}{\alpha}$$

we obtain

$$d\Upsilon = \frac{1}{\alpha}\, di,$$

so

$$\tau = \frac{1}{\alpha}\int \varepsilon\, di.$$

Time is thus ε-weighted memory writing, scaled by parsimony. This formulation also makes the arrow of time immediate, because if the ratchet enforces $d\Upsilon \geq 0$ and $\varepsilon \geq 0$, then

$$d\tau \geq 0.$$

Entropy polarity becomes therefore becomes temporal direction without additional assumption.

Thermodynamic Conjugacy

To make the structure explicit, introduce a potential $G(\Upsilon)$, representing the minimal free-energy cost required to sustain a given history depth. Define

$$\varepsilon \equiv \frac{\partial G}{\partial \Upsilon}.$$

Then

$$d\tau = \varepsilon\, d\Upsilon = dG,$$

and along any realised trajectory,

$$\tau = \tau_0 + \Delta G.$$

Time therefore equals the accumulated energetic cost of committed structure. This identifies ε as a costate variable, conjugate to Υ. The pair (Υ, ε) functions analogously to thermodynamic conjugates, but with history as the macroscopic coordinate and energy cost as its conjugate effort.

Dilation as Cost Variation

Consider two regions that accumulate the same increment $d\Upsilon$, but with different marginal costs ε_1 and ε_2:

$$d\tau_1 = \varepsilon_1 d\Upsilon, d\tau_2 = \varepsilon_2 d\Upsilon.$$

Then

$$\frac{d\tau_1}{d\tau_2} = \frac{\varepsilon_1}{\varepsilon_2}.$$

Time dilation arises from variation in the cost of maintaining history, and geometry later encodes this variation. The underlying mechanism is energetic.

The Λ Ceiling

History depth is globally bounded:

$$\Upsilon \leq \Upsilon_{\max}(\Lambda).$$

Therefore total emergent time is bounded by

$$\tau_{\max}(\Lambda) = \tau_0 + \int_0^{\Upsilon_{\max}(\Lambda)} \varepsilon(\Upsilon)\, d\Upsilon.$$

The cosmological constant therefore sets a ceiling on how much history can exist, and consequently on how much time can be generated. Local scaling efficiency is governed by α, while global capacity is governed by Λ. Time then measures the energetic accumulation of committed history within these bounds.

Relation to Einstein

In General Relativity, proper time is determined by the spacetime metric. Clocks tick at rates set by geometry. In the DE framework, geometry is secondary. Time is generated through the cost of sustaining history, and when geometry becomes the most economical bookkeeping surface, the marginal cost ε is encoded in the metric component g_{00}. Proper time is therefore the geometric representation of accumulated history cost.

Once sufficient history has been accumulated and mass becomes abundant, gravity emerges and geometry stabilises as the least-cost encoding of constraint. In that regime,

$$d\tau = \sqrt{g_{00}}\, dt \;\approx\; \left(1+\frac{\Phi}{c^2}\right) dt,$$

while in DE,

$$\frac{d\tau_1}{d\tau_2} = \frac{\varepsilon_1}{\varepsilon_2}.$$

These correspond when

$$\varepsilon \propto \sqrt{g_{00}}.$$

In this sense, ε acts as the non-geometric precursor of the metric time component. Geometry encodes what recursion has already stabilised, while time records the accumulated irreversible cost of that stabilisation. Relativity then describes how this cost is distributed once geometry has emerged.

Feigenbaum Scaling

Each renormalisation step writes history, each sacrifice of freedom commits constraint, and scaling accumulates cost.

$$d\tau = \varepsilon \, d\Upsilon.$$

Feigenbaum's α governs how efficiently that transition propagates locally. The cosmological constant Λ sets the global capacity within which it unfolds, and time measures the energetic accumulation between them. Divergences reflect unresolved scale transitions, horizons reflect capacity limits, and curvature reflects accumulated commitment. Renormalisation is the geometry of viable scaling and it marks the boundary where uncertainty becomes structure.

In the DE framework renormalisation is interpreted as a bifurcation event in the effective dynamics of a system. When the cost of maintaining microscopic freedom exceeds the cost of committing higher-order structure, the system reorganises. Degrees of freedom collapse into a new attractor basin and a new coarse-grained description becomes valid. Renormalisation therefore marks the point at which a system crosses a stability threshold and its dynamical landscape changes. Feigenbaum scaling suggests that viable recursive systems may converge toward a characteristic survival fraction close to $1/\alpha$, balancing exploratory freedom against structural inheritance. This predicts that many biological and regulatory transitions should show an effective dimensional contraction near 0.4 at collapse. Renormalisation corresponds to increments in accumulated memory that stabilise under the contraction–rotation operator Υ. At each step, a subset of correlations is preserved and promoted into committed structure, increasing the depth of memory

$$\Delta\Upsilon \propto \frac{\Delta i}{\alpha}$$

where α encodes the minimal contraction required for stability. In this way, Υ acts as the ratchet that converts informational possibility into structural persistence. From this perspective, structure is seen as accumulated history. A molecule retains the constraints of its atomic interactions. A cell encodes the chemistry from which it emerged. At every scale, the system carries forward a compressed record of its prior states through successive renormalisation steps. The universe therefore does not merely evolve, it retains memory.

This interpretation introduces a natural distinction between attractors. Not all attractors support continued recursion. A viable or "true" attractor is one that preserves sufficient structured correlation to allow further renormalisation. In DE terms, such an attractor maintains a positive entropic flow along the local orientation,

$$\frac{d\Upsilon}{d\varepsilon} \geq 0$$

while retaining enough residual degrees of freedom for subsequent bifurcations. It stabilises correlations without exhausting the system's capacity for further emergence. Renormalisation thus marks the boundary at which possibility becomes commitment. It is the point at which the system selects a direction along the ε axis and encodes that selection into Υ. What survives becomes the substrate for further recursion, and what is discarded is removed from the accessible state space.

In this formulation, complexity emerges as a cumulative effect of successive ϒ-driven stabilisations. Each level of organisation exists because the previous level preserved just enough structure to support it. The $\varepsilon - \Upsilon$ recursion therefore provides a unified description of how systems move from uncertainty to structure across scales. If renormalisation is the process that creates memory, then the evolution of that memory must follow a coherent law. In the DE framework, that law takes the form of an oscillator in entropic distance,

$$\frac{d^2\Upsilon}{d\varepsilon^2} = -\Omega^2\Upsilon$$

with ε defining the orientation of motion and ϒ governing the stabilisation of its cycles. The universe can then be understood as a recursive system that alternates between exploration and commitment, generating structure through repeated transitions across the same fundamental geometry. In the classical period-doubling cascade:

- **Feigenbaum alpha (~2.5029..)**
  → controls **geometric scaling in parameter space**
  → how quickly bifurcation points accumulate

- **Feigenbaum δ (~4.669...)**
  → controls **scaling of orbit amplitudes in phase space**

→ how distances between points in the attractor shrink across levels

So:

- α = scaling of when bifurcations occur
- δ = scaling of what the system looks like after each bifurcation

If we then consider that each Feigenbaum bifurcation as a renormalisation step (ϒ operator), where each step compresses degrees of freedom ($\sim 1/\alpha \approx 0.4$), and collapse occurs when uncertainty cost exceeds compression benefit, then δ governs how the internal structure reorganises after each compression.

$\delta \approx$ scaling of residual structure after compression

After each ϒ (compression) α tells us how much of the system survives, then δ tells us how that surviving structure is redistributed. We can think of δ as controlling the internal geometry of memory, so that the attractor is not just smaller, it is reshaped and δ determines how phase-space distances contract unevenly. In effect δ governs memory density redistribution along the *i-axis*. This influences the stability of the emergent attractor, where larger δ-values produces more uneven contraction with sharper structure, whilst smaller δ-values leads to smoother contraction, and a more diffuse attractor

Predictive granularity

Each renormalisation step reduces degrees of freedom (α) reweights what remains (δ), so that we can think of α = quantity of retained structure and δ = quality / arrangement of retained structure. Collapse occurs when added degrees of freedom increase uncertainty cost, so δ enters here because if δ produces highly uneven contraction, you get a tightly bound attractor, with strong correlations and a faster approach to constraint saturation. If δ is more uniform, however, then the system tolerates more iterations before collapse. So δ modulates how quickly the system reaches critical threshold.

Therefore in the DE framework, renormalisation events correspond to bifurcation steps in which approximately $\frac{1}{\alpha}$ of accessible degrees of freedom are compressed into invariant structure. The Feigenbaum constant δ governs the internal rescaling of this surviving structure, determining how correlations and phase relationships are redistributed across iterations. While α sets the rate of entropy reduction, δ defines the geometry of retained memory, thereby modulating the approach to constraint saturation and the onset of collapse. Formally;

1. Start from the DE renormalisation step

Let a system at level n have:

- accessible degrees of freedom: D_n
- retained invariant structure: I_n
- memory load or unresolved constraint: M_n
- correlation density: C_n

Your renormalisation operator υ maps

$$\Upsilon: (D_n, I_n, M_n, C_n) \mapsto (D_{n+1}, I_{n+1}, M_{n+1}, C_{n+1})$$

with the first-order DE compression rule

$$D_{n+1} \approx \frac{D_n}{\alpha}$$

so $\frac{1}{\alpha} \approx 0.4$ is the fraction retained in compressed form. That is the global shedding rule.

2. Where δ enters

In the standard bifurcation picture, α governs the rescaling of distances in state space near the critical attractor, while δ governs the convergence of bifurcation intervals in control-parameter space;

$$\delta = \lim_{n\to\infty} \frac{\Delta\lambda_n}{\Delta\lambda_{n+1}}$$

where $\Delta\lambda_n$ is the spacing between successive bifurcation values. So after each renormalisation step, the surviving structure is not merely smaller, it is reshaped, with some correlations pulled closer together, some phase relations sharpened, some regions effectively abandoned.

So the DE reading is:

$$I_{n+1} = \mathcal{R}_\delta(I_n)$$

where $\mathcal{R}_\delta$ is a selective geometric redistribution.

3. Mapping δ to the *i-axis*

If the *i-axis* is the carrier of predictive memory, then δ is naturally interpreted as controlling the density gradient of retained memory along that axis;

$$\rho_{i,n+1}(x) = \delta\text{-weighted redistribution of } \rho_{i,n}(x)$$

where $\rho_i(x)$ is memory density along the internal axis.

4. Holonomy interpretation

If memory is stored through holonomy on the $U(1)$ fibre, then each renormalisation step does two things:

1. compresses accessible microstructure
2. increases the curvature burden carried by what remains

So δ can be interpreted as controlling the rate at which curvature becomes concentrated in the retained structure:

$$\Omega_{n+1} \sim \delta\,\Omega_n$$

not necessarily as a literal scalar multiplication everywhere, but as the statement that δ governs the scaling of effective curvature concentration across recursive levels. Then collapse occurs when accumulated holonomy can no longer be smoothly distributed;

$$\sum_{k=0}^{n} \Delta\,\theta_k \to \theta_{\text{crit}}$$

with $\Delta\theta_k$ shaped by δ.

Thus α determines the rate of coarse-graining, while δ determines the geometric concentration of phase-memory that pushes the system toward threshold.

5. Threshold condition

Define uncertainty cost:

$$U_n = f(M_n, C_n, \Omega_n)$$

Then after each renormalisation:

$$U_{n+1} = f\,(M_{n+1}, C_{n+1}, \mathcal{R}_\delta(\Omega_n))$$

Collapse occurs when

$$\Delta U_{n+1} > \Delta \Pi_{n+1}$$

where Π_{n+1} is the predictive gain from further compression. If another renormalisation step therefore gives more predictive simplification than structural cost, the system remains metastable, but if δ-driven concentration makes curvature and correlation costs rise faster than predictive gain, the system collapses into an entangled attractor. So in this frame δ is one of the main determinants of how quickly the cost curve steepens. What survives gets folded back into the system in a particular shape. That shape matters because collapse is never caused by raw quantity alone, rather it is caused by concentrated unresolved structure.

This also helps explain why the emergent state is not the sum of the parts. After enough renormalisation steps the parts are no longer represented independently as δ has recursively rescaled their relations. Memory becomes concentrated in invariant phase structure and the system crosses into a regime where only the attractor is predictive. So the entangled entity becomes a δ-shaped memory object;

Effective dimensionality contracts under recursive renormalisation. Each recursion step reduces the number of independent accessible degrees of freedom by the

Feigenbaum contraction factor α, progressively compressing uncertainty into structured constraint.

$$I_{n+1} = \mathcal{R}_\delta(I_n)$$

Retained informational structure is recursively redistributed through the nonlinear renormalisation operator $\mathcal{R}_\delta$. The Feigenbaum scaling constant δgoverns how surviving correlations are reorganised and concentrated across scales.

$$\Omega_{n+1} = \Phi_\delta(\Omega_n, M_n, C_n)$$

The system state evolves through nonlinear recursive transformation. Future structure depends on the prior attractor geometry Ω_n, retained memory M_n, and accumulated constraint C_n, reshaped under δ-governed redistribution.

$$\text{Collapse occurs when } U_{n+1} > \Pi_{n+1}$$

Collapse is triggered when unresolved uncertainty Uexceeds the system's remaining predictive or stabilising capacity Π. Beyond this threshold, the system can no longer sustain competing configurations and must commit to a stable attractor state.

with:

- D_n: accessible degrees of freedom
- I_n: retained invariant structure
- Ω_n: effective curvature or holonomy burden
- U_n: uncertainty cost
- Π_n: predictive compression gain

If we only had compression, the system would just shrink, lose detail, and nothing new would emerge. The sigma variable changes that by reorganising what survives, it builds structure. Over many steps, this creates tightly connected patterns that behave as single entities rather than loose collections of parts, and this is where emergence comes from. At some point, the system becomes so tightly organised that it can no longer be understood by looking at its pieces separately. It has effectively become something new, a unified structure with its own behaviour, and that is the moment where the system becomes effectively entangled. Thus δ is the scaling law of correlation concentration and entanglement is the state where that concentration saturates.

Because δ contracts distances unevenly it creates regions of very high correlation. That uneven contraction is exactly what produces clustering, locking of phase relationships, and dependency between components, which are the hallmarks of entanglement-like behaviour. Entanglement is the regime in which that concentration becomes irreducible, while δ defines the constraint topography that makes entanglement the inevitable attractor. If δ defines the topography, entanglement is selected by the system because as the only stable configuration remaining.

The transition from stochastic to deterministic emergence occurs when δ-driven concentration of relational structure outpaces α-driven reduction in accessible degrees of freedom, causing the effective dimensionality of the system to collapse toward a single predictive manifold. The same structure can be mapped across models by treating:

- α as the rate at which possibilities are reduced

- δ as the rate at which retained structure becomes correlated
- $\mathcal{P}$ as the degree to which higher-level behaviour becomes predictable

For any system:

$$\mathcal{P}_n = 1 - \frac{D_n^{\text{eff}}}{D_0}$$

where

$$D_n^{\text{eff}} = \frac{D_n}{\kappa_n}$$

and

- D_n= remaining accessible possibilities
- κ_n= concentration of correlation, memory, or coupling
- $\mathcal{P}_n$= predictability of the emergent state

Thus a system becomes more predictable when it has fewer independent options and stronger internal coupling. Rising predictability means the system is losing independent futures and becoming governed by a smaller set of tightly coupled constraints. The DE model predicts that systems become more forecastable not simply when information accumulates, but when independent possibilities collapse into a smaller number of correlated futures. From this we can create a universal model in which systems evolve by losing independent degrees of freedom and concentrating the remainder into correlated structure, with predictability emerging as a consequence of that process. Predictability is the shadow cast by lost independence.

Appendix.

Thus we can formalise a predictability spectrum driven by δ and the approach to threshold;

1. Define what we mean by predictability

We separate two things:

- **Structural predictability**
  → can we predict the type of outcome (e.g. collapse, attractor, phase locking)

- **State predictability**
  → can we predict the exact configuration

The framework already implies structural predictability rises faster than state predictability

2. Core variables

Let the system at level n have:

- D_n: degrees of freedom
- C_n: correlation density
- Ω_n: curvature / holonomy burden
- U_n: uncertainty cost
- Π_n: predictive gain
- δ: internal scaling factor

3. δ-driven concentration

Define a concentration function:

$$\kappa_n = g(\delta, n)$$

This represents how strongly structure has been compressed into correlated form. So:

$$\kappa_n \propto \delta^n$$

Interpretation:

- early iterations → low concentration
- later iterations → rapid clustering of correlations

4. Effective degrees of freedom

Not all degrees of freedom remain independent:

$$D_n^{\text{eff}} = \frac{D_n}{\kappa_n}$$

So:

- α reduces D_n globally
- δ reduces independence within D_n

5. Predictability index

Now define a dimensionless predictability index:

$$\mathcal{P}_n = 1 - \frac{D_n^{\text{eff}}}{D_0}$$

So:

- $\mathcal{P}_n \approx 0$→ highly unpredictable

- $\mathcal{P}_n \to 1 \to$ highly constrained, near-deterministic

Substitute:

$$\mathcal{P}_n = 1 - \frac{D_n}{D_0\, \kappa_n}$$

Using DE scaling:

$$D_n \sim \frac{D_0}{\alpha^n}$$

So:

$$\mathcal{P}_n = 1 - \frac{1}{\alpha^n\, \kappa_n}$$

If $\kappa_n \sim \sigma^n$, then:

$$\mathcal{P}_n = 1 - \left(\frac{1}{\alpha\delta}\right)^n$$

6. Result

If $\alpha\delta > 1$, then

$$\mathcal{P}_n \to 1$$

predictability rises with recursion

- if $\alpha\delta \approx 1$, system stays marginal
- if $\alpha\delta < 1$, predictability does not accumulate

Since empirically:

- $\alpha \approx 2.5$
- $\delta \approx 2.7$

Then:

$$\alpha\delta \approx 6.7 \gg 1$$

Thus predictability should increase rapidly across renormalisation levels, where predictability grows exponentially with recursion whenever α and δ jointly exceed unity.

7. Link to collapse threshold

Collapse occurs when:

$$U_n > \Pi_n$$

But:

- U_n increases with κ_n (concentration → higher constraint cost)
- Π_n decreases as D_n shrinks

So threshold occurs roughly when:

$$\kappa_n \sim D_n$$

Substitute:

$$\delta^n \sim \frac{D_0}{\alpha^n}$$

Which gives a critical n^*:

$$(\alpha\delta)^n \sim D_0$$

So collapse occurs when δ-driven concentration overtakes α-driven reduction

8. Interpretation of the spectrum

We now get three regimes:

(1) Early regime: low predictability

- high D_n, low κ_n
- many independent paths
- emergence weak and variable

(2) Intermediate regime: structured emergence

- α reducing options
- δ clustering structure
- attractors begin forming
- structural predictability high

(3) Near-threshold regime: deterministic emergence

- $D_n^{\text{eff}} \to 0$
- correlations dominate
- only a few viable configurations remain

→ system behaves as a single entity. Predictability increases as the ratio of δ-driven correlation concentration to α-driven degree-of-freedom reduction grows. When this ratio exceeds unity across recursive levels, the system transitions from high-variance behaviour to constrained, near-deterministic emergence.

Chapter 9: The Harmonic Oscillator

In General Relativity, the relation between matter and geometry is expressed through Einstein's equation

$$G_{\mu\nu} = 8\pi\, T_{\mu\nu}$$

where the stress–energy tensor $T_{\mu\nu}$ describes the distribution of matter and energy, and the Einstein tensor $G_{\mu\nu}$ describes the curvature of spacetime. In the Deductive Entropy framework, this relation must emerge from the underlying recursion that generates structure. The primary coordinate of this process is the ordering parameter ε, which tracks progression through the recursion. This progression alone does not produce structure. At each step, correlations are filtered by the stabilisation operator Υ, which determines what can persist under continued recursion.

However, Υ acts locally. It determines what survives at a given step, but it does not by itself provide a record of what has been retained. Without such a record, there is no distinction between transient survival and persistent structure. We therefore require a quantity that accumulates the retained portion of the recursion. The minimal definition of such a quantity is

$$\frac{d\Sigma}{d\varepsilon} = \Upsilon(\varepsilon)$$

so that

$$\Sigma(\varepsilon) = \int_0^{\varepsilon} \Upsilon(\varepsilon')\, d\varepsilon'$$

Here, Υ determines what survives locally, while Σ records what has survived globally. Structure is therefore not generated by ε alone, but by the accumulated effect of stabilisation across recursion. Once defined, Σ provides a measure of retained structural memory. Regions in which Σ becomes large correspond to configurations that have repeatedly survived stabilisation and therefore strongly resist perturbation. In this sense persistence is built up through recursion. If Σ varies across space, then differences in retained structural memory appear,

$$\nabla_\mu \Sigma \neq 0$$

and these gradients quantify how difficult it is for the system to transition between neighbouring configurations. Motion then follows paths that minimise variation in retained structural memory, guided by the local orientation of ε. In the continuum limit, such paths correspond to geodesics. Geometry therefore does not precede structure, but emerges from the spatial distribution of retained structural memory. Regions of high accumulation produce strong gradients, which manifest as resistance to change and the familiar effects associated with mass.

Only at this stage does it become meaningful to compare with General Relativity. The stress–energy tensor can be written as an effective functional of Σ,

$$T_{\mu\nu}^{(\Sigma)} = \nabla_\mu \Sigma \nabla_\nu \Sigma - g_{\mu\nu}\left(\frac{1}{2}\nabla^\alpha \Sigma \nabla_\alpha \Sigma + V(\Sigma)\right)$$

where the kinetic term describes the flow of retained structure and the potential $V(\Sigma)$ represents stored structural curvature. Einstein's equation then takes the form

$$G_{\mu\nu} = 8\pi\, T_{\mu\nu}^{(\Sigma)}$$

indicating that spacetime curvature is the geometric response to the distribution of retained structural memory.

The remaining question is dynamical. If Σ encodes accumulated structure, its evolution must balance continued accumulation against destabilisation. The simplest closed form that captures this balance is harmonic,

$$\frac{d^2\Sigma}{d\varepsilon^2} = -\Omega^2\Sigma$$

which describes oscillation about stable configurations in the space of retained structure. In this formulation, ε governs progression, Υ governs local selection, and Σ records accumulated persistence. Gravitational behaviour then emerges as the large-scale consequence of how this retained structural memory is distributed and evolves.

Figure: Chiral structure under recursive emergence

In this double helix image the vertical axis aligns with the local $U(1)$ phase direction (*i-axis*), parameterised by the recursion coordinate ε. The intertwined helices represent a chiral doublet undergoing $SU(2)$ rotations under recursive transport; their winding encodes phase progression and the accumulation of retained structure $\Sigma(\varepsilon)$ under stabilisation Υ. Colour variation along the strands represents $SU(3)$ internal degrees of freedom (colour space), indicating constrained transformations within the strong sector. The surrounding shells depict constraint envelopes, regions of admissible transport, suggesting a holographic-like encoding in which boundary constraints reflect interior coherence. Stable structure arises where transport, phase alignment, and internal rotations remain mutually consistent across recursion, while loss of coherence corresponds to regions where accumulated structure cannot be sustained. The two intertwined helices represent the dual chiral modes ψ_L and ψ_R, whose relative orientation encodes phase and internal symmetry. Their continuous winding reflects recursive transport and the accumulation of retained structure, corresponding to $\Sigma(\varepsilon)$. The local pitch and coherence of the helices indicate regions where stabilisation Υ successfully preserves correlation across steps.

The same recursive process that stabilises chiral degrees of freedom operates at all scales. What appears as gauge structure in the microscopic domain becomes large-scale organisation when extended across the full recursion. The surrounding shells in the figure represent constraint surfaces, regions within which transport remains consistent and memory can be preserved. These define boundaries across which information must be encoded rather than transmitted directly. In this sense, structure is thus constrained by the surfaces that bound it.

This establishes a natural holographic correspondence. The state of a region is determined not only by its internal configuration, but by the constraints imposed at its boundary. Memory is therefore distributed across surfaces of consistency rather than propagated through space. The interior reflects the accumulated effect of boundary constraints under recursion. When extended cosmologically, this implies that regions that appear distant in spacetime may nevertheless share a common constraint surface in ε. Their apparent separation is geometric, but their origin is informationally coupled through the same ordering process. Large-scale coherence therefore does not require causal contact in spacetime, because it arises from shared participation in the same recursion domain.

In this way, holography provides the bridge between local stabilisation and global structure. The same mechanism that enforces consistency across gauge sectors also enforces coherence across the universe. The distinction between microstructure and cosmology becomes one of scale, not of principle.

In standard cosmology, inflation is introduced to explain why distant regions look similar. However uniformity can be also explained as emergence from a shared entropic interface. Regions that appear far apart geometrically originate from the same ε ordering domain, so that large-scale isotropy is reframed as input from shared origin. Hubble deep-field observations sees that earlier galaxies are smaller, denser, more irregular, star formation rates peak then decline, and structure becomes more hierarchical over redshift.

This matches DE expectations exactly, where early ε shallow basins produce noisy, rapidly assembling structure, with later deeper attractors producing stable galaxies with clearer morphology. Redshift tracks ordering depth, because it measures how far a structure is from the current recursion state, not how far it is from a singular beginning. In this way all standard observables can be recovered as effective descriptions of the same ε ordering dynamics, where the high temperatures of the early universe arise from rapid entropic resolution prior to deep attractor formation, and large-scale uniformity reflects a shared entropic origin rather than causal equilibration within spacetime. Heat is accounted for as unresolved entropic excitation, cooling follows from stabilisation and Hubble observations align naturally with hierarchical basin deepening. In this way inflationary signatures are reproduced without a Big Bang singularity.

The Chiral Waveform is Born with Phase π at the SU2 Bifurcation

The constant π appears throughout physics as a consequence of rotational symmetry, phase periodicity, and compact group structure. In all standard treatments, it is a geometric constant. In the DE framework, it plays a deeper role because it marks the first bifurcation in phase that enables relational structure. In the SU(2) regime, recursion produces a two-phase structure with a fixed separation of π. This phase offset defines chirality. The system splits into two branches, ψ_L and ψ_R, whose relative orientation encodes direction. Transient correlations form under recursive contraction and rotation, and structure is leaky. Stability requires closure and this occurs when the recursion admits a third branch, corresponding to the $\frac{2\pi}{3}$ phase structure of SU(3). At this point, the two-phase system can embed within a triadic configuration. The geometry closes, correlations can no longer dissipate freely, and a stable attractor appears. This transition marks the conversion of correlation into persistent entanglement. It is mediated by the stabilisation operator Υ, which selects configurations that remain consistent across recursion. The mapping can be expressed schematically as

$$\text{Correlation}_{SU(2)} \longrightarrow \text{Entanglement}_{SU(3)}$$

Once this transition occurs information is no longer transient, and it becomes encoded in stable relational structure. In the Standard Model, this appears as confinement and gluon self-interaction, which enforce the persistence of colour configurations. In DE terms, this is the point at which the system begins to retain structure rather than merely explore it. The role of π in this process is fundamental, as it defines the initial phase separation that produces chirality, and therefore the first non-trivial relational degree of freedom. It is a bifurcation parameter that determines when symmetry can break into structured orientation. In this sense, π acts as a transition constant, and it marks the boundary between undifferentiated phase and directed structure, enabling the subsequent emergence of triadic stability and persistent entanglement. The stabilisation operator Υ then acts as a ratchet, promoting only those configurations in which this structure can be maintained across recursion.

The Schwarzschild horizon as a chiral pivot

This graph, taken from Einstein's equations, shows how time slows to a halt at the event horizon of a black hole. The golden graph traces a single quantity taken straight from Einstein's general theory of relativity, the time factor in the Schwarzschild metric:

$$T(r) = \sqrt{1 - \frac{2GM}{r}}.$$

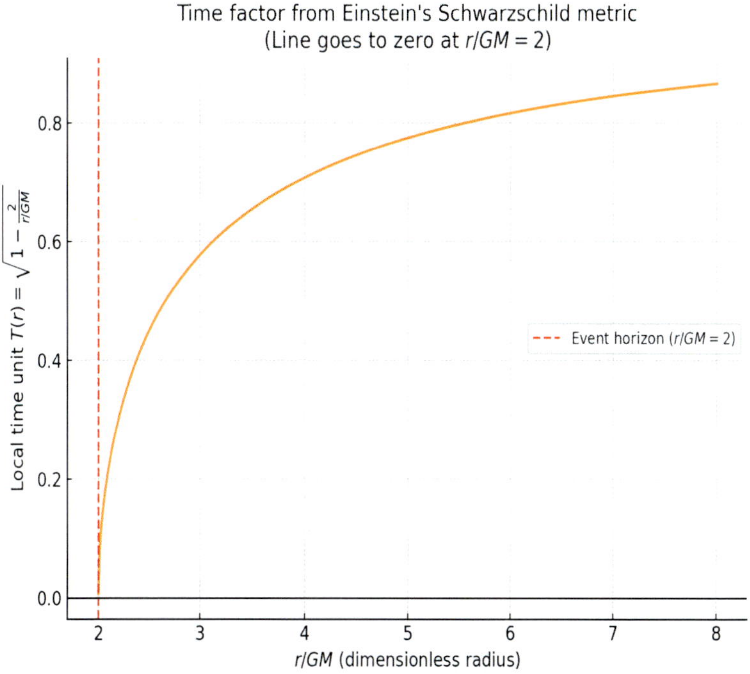

It describes how local clock rates vary with radius in a gravitational field. As $r \to 2GM$, the factor approaches zero, and time dilation becomes infinite for a distant observer. In standard General Relativity, this marks the event horizon, a causal boundary beyond which signals cannot escape. However, the mathematical structure contains more than a limiting value. The square root implicitly admits two branches:

$$\varepsilon_{\pm}(r) = \pm\sqrt{1 - \frac{2GM}{r}}$$

The positive branch corresponds to the conventional interpretation of forward temporal progression. The negative branch is normally discarded as unphysical. In the DE framework, this second branch is essential. At $r = 2GM$, both branches meet at zero. This is a bifurcation point where the orientation of emergence becomes undefined and can reverse. The horizon is therefore a chiral pivot.

On one side, $\varepsilon > 0$, corresponding to the observed universe, where entropy increases and structure unfolds forward. On the other, $\varepsilon < 0$, a conjugate branch in which the direction of emergence is inverted. The two branches are complementary phases of a single structure. This reinterpretation resolves the apparent singular behaviour of the Schwarzschild metric. The vanishing of the time factor is the point at which the system transitions between chiral modes. Information does not halt at the horizon, it changes orientation.

If the two branches are plotted together, they form a symmetric pair about $\varepsilon = 0$. The resulting structure has the defining features of a harmonic system, with two conjugate states, a shared turning point, and a conserved quantity that moves between them without loss. The horizon becomes the node of this system, where accumulated structure is exchanged between branches. What appears as gravitational time dilation is the projection of a more fundamental process, the bifurcation of emergence into chiral modes. The event horizon is thus the interface through which this bifurcation is continuously realised.

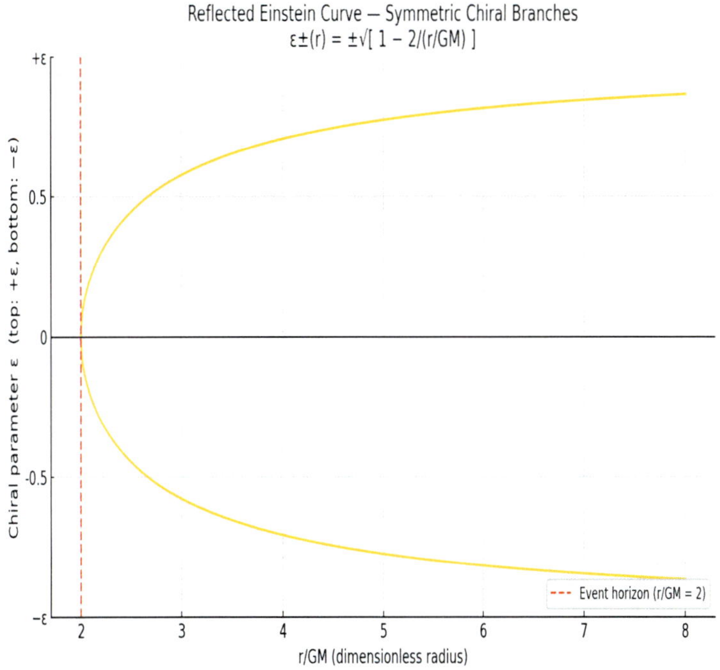

Instead of describing how time slows, it now reveals how chirality splits. At the event horizon, $r/GM = 2$, both branches meet at zero. This is a bifurcation point in emergence itself, the point at which the direction of informational flow can reverse. On one side, $\varepsilon > 0$, corresponding to the observable universe where entropy increases and temporal ordering appears forward-directed. On the other, $\varepsilon < 0$, a conjugate branch in which the orientation of emergence is inverted. These are complementary phases of a single structure, and the reflected curve makes this explicit:

$$\varepsilon_{\pm}(r) = \pm\sqrt{1 - \frac{2}{r/GM}}$$

The horizon becomes the node at which these branches meet. This carries several immediate consequences. The horizon acts as a bifurcation point, analogous to the centre of a Feigenbaum diagram, where a single solution splits into two stable branches. The system divides into conjugate modes of existence, whilst the two branches together form a continuous structure. Information does not vanish at the horizon, it is conserved through inversion. The flow of emergence reverses sign, but the underlying structure remains intact. In this sense, the horizon can be seen as an interface.

This symmetry also unifies what are conventionally treated as distinct regions. The interior and exterior of a black hole become two sides of the same entropic surface. The reflected branches form a standing structure, with one phase advancing as the other recedes, each sustaining the other's evolution. Seen this way, the bifurcation is not an event in the past, it is the ongoing condition of the universe. Temporal ordering arises locally from the gradient of ε, rather than from a universal clock, and what appears as cosmic expansion reflects the continuous separation of these conjugate branches. This reframes several foundational ideas:

- The Big Bang is not a singular origin, but a persistent interface.
- Time is not fundamental, but emerges from the orientation of the entropic gradient.
- The universe exists as paired chiral branches, continuously exchanging informational flux.

- Apparent vacuum energy reflects the coupling between these branches.

The reflected structure has the defining features of a harmonic system, with two conjugate branches, a shared turning point, and a conserved quantity that transfers between them. However, this is not an oscillation in space or time, but in the orientation of emergence itself. The shared zero at $\frac{r}{GM} = 2$ acts as the pivot at which the sign of εreverses, analogous to the turning point of an oscillator where motion changes direction while continuity is preserved. In this formulation, gravitational horizons are not boundaries but switching surfaces of a two phase oscillator. What appears as a causal limit in General Relativity becomes, in DE, the interface across which emergence changes orientation. Thus the universe maintains structure by continuously transporting information across the divide.

Planck's Constant as the Minimum Entropic Distance

Phase is the primary variable of emergence. It encodes ordering and orientation before any notion of persistent structure. What is retained as structure is the portion of phase progression that survives stabilisation. This accumulated retention has already been defined as

$$\Sigma(\varepsilon) = \int_0^{\varepsilon} \Upsilon(\varepsilon')\, d\varepsilon'$$

Because Σ accumulates stabilised phase, it carries the dimensions of action. To render this phase observable, a scale is required. That scale is Planck's constant, and the observable phase is therefore

$$\theta = \frac{\Sigma}{\hbar}$$

Planck's constant sets the minimum increment of retained phase required to distinguish one configuration from another. When

$$\Delta\Sigma \ll \hbar$$

phase differences are unresolved and neighbouring recursion states are indistinguishable. Only when

$$\Delta\Sigma \sim \hbar$$

does a distinct configuration emerge. Planck's constant therefore defines the minimum resolvable increment of retained structure,

$$\Delta\Sigma_{\text{min}} = \hbar$$

Quantisation follows from closure. For a closed recursion loop, phase must be single-valued,

$$\oint d\theta = 2\pi n \Rightarrow \oint d\Sigma = 2\pi n\hbar$$

Stable structures arise only when retained phase accumulates in discrete units. Classical behaviour emerges in the limit $\Sigma \gg \hbar$, where rapidly varying phase cancels except at stationary points. In this formulation, quantum mechanics quantises the accumulation of stabilised phase.

Phase Subdivision and the Resolution of Chirality

Because phase is primary, the constant π acquires direct ontological meaning. A chirality inversion corresponds to a phase shift

$$\Delta\theta = \pi \Rightarrow \Delta\Sigma = \pi\hbar$$

This is the minimal bifurcation in phase space, the smallest separation that produces a distinct orientation of emergence. In the SU(2) regime, recursion separates into two chiral branches, ψ_L and ψ_R, with a fixed phase offset of π. This establishes direction but does not yet stabilise structure. Correlations remain transient because phase is not closed. Stability emerges only when the system admits a triadic closure, corresponding to the SU(3) regime. At this point, the stabilisation operator ϒ confines phase into a closed relational structure. Correlations can no longer dissipate, and persistent entanglement appears. The retained result accumulates as Σ. Structure therefore arises when phase is no longer free to drift, but is confined into a closed configuration that preserves itself across recursion.

Once π has defined chiral opposition and $\frac{2\pi}{3}$ has defined triadic closure, the remaining phase structure is no longer arbitrary. The system cannot transition directly from opposition to full closure without resolving intermediate structure. The allowed phase intervals are therefore constrained by the need to preserve conjugacy, rotational closure, and stability under recursive compression. The first stable subdivision is:

$$\frac{\pi}{3}$$

This is the smallest phase spacing that preserves a sixfold conjugate structure: three orientations and their three mirror partners. It is the minimal interval compatible with triadic closure. At this scale, phase relationships can cycle without loss, and structure becomes self-consistent under recursion.

This marks the first fully stable chiral configuration. A further subdivision gives:

$$\frac{\pi}{6}$$

This is the half-step between neutrality and full sixfold locking. It is the earliest resolvable bias within an already chiral system. At this level, the system exhibits directional preference in ε, but has not yet fully committed to triadic closure. It represents a metastable tilt that can either collapse or be amplified depending on the local entropic conditions. The hierarchy is therefore:

$$\pi \rightarrow \frac{2\pi}{3} \rightarrow \frac{\pi}{3} \rightarrow \frac{\pi}{6}$$

or, expressed physically:

opposition → closure → stable conjugate spacing → early directional bias

This sequence emerges from the constraints of phase continuity and recursive stabilisation. π establishes directional opposition in ε and triadic closure enforces confinement. The subsequent subdivisions describe how the system resolves this structure into stable and metastable configurations. Chirality originates at π and the smaller intervals describe how that initial bifurcation becomes structured, stabilised, and ultimately irreversible under recursion.

The Harmonic Oscillator Structure

The DE recursion therefore behaves as a pair of coupled oscillators, with one governed by the entropy orientation ε one governed by the contraction-rotation operator Υ, and together they form the universal recursion pair

$$\Xi = (\varepsilon, \Upsilon)$$

This pair evolves along the complex phase axis (the U(1) fibre), producing a standing oscillatory pattern across recursion depth. The system therefore alternates between two phases separated by π at the SU2 bifurcation. The resulting waveform is a standing oscillation in the topological axes of the system and this oscillation produces alternating dominance of the two conjugate phases while maintaining global coherence. These two variables drive the recursion where ε determines orientation, and Υ determines stabilisation.

The oscillator thus produces a ratchet, because every time correlations stabilise they cannot simply dissolve back into noise without breaking the symmetry that created them. This creates successive domains of stability, where each domain corresponds to a level at which correlations can survive longer than the noise acting on them. Complexity therefore grows stepwise rather than continuously, and once stable entanglement exists the DE recursion begins to climb a hierarchy of domains. Each domain corresponds to a new level at which correlations survive perturbation and feedback on themselves. The sequence follows a simple rule, so that when correlations persist long enough they become structure, and the hierarchy unfolds as successive stability domains;

At every level the same recursion operates, only the physical expression changes. Structure persists when correlations cannot be factorised. Particles, atoms, molecules, and organisms persist because their correlations form closed entanglement networks. Fields act as buffering layers within this system, as they distribute interaction, absorb perturbation, and allow correlations to reorganise without collapse. Structure therefore follows three simple rules where entanglement density determines stability, entanglement topology determines function, and entanglement persistence determines time. Where correlations remain closed and buffered, structure persists, and where they disperse, structure dissolves.

Quantum mechanics therefore governs structure wherever correlations remain coherent. It is not restricted to microscopic systems, rather the classical world is simply the regime in which correlations become sparse enough that quantum coherence no longer survives. Quantum mechanics therefore is the physics of persistent correlation, and wherever entanglement remains dense, recursive, and protected against decoherence, the same stability rules apply regardless of scale.

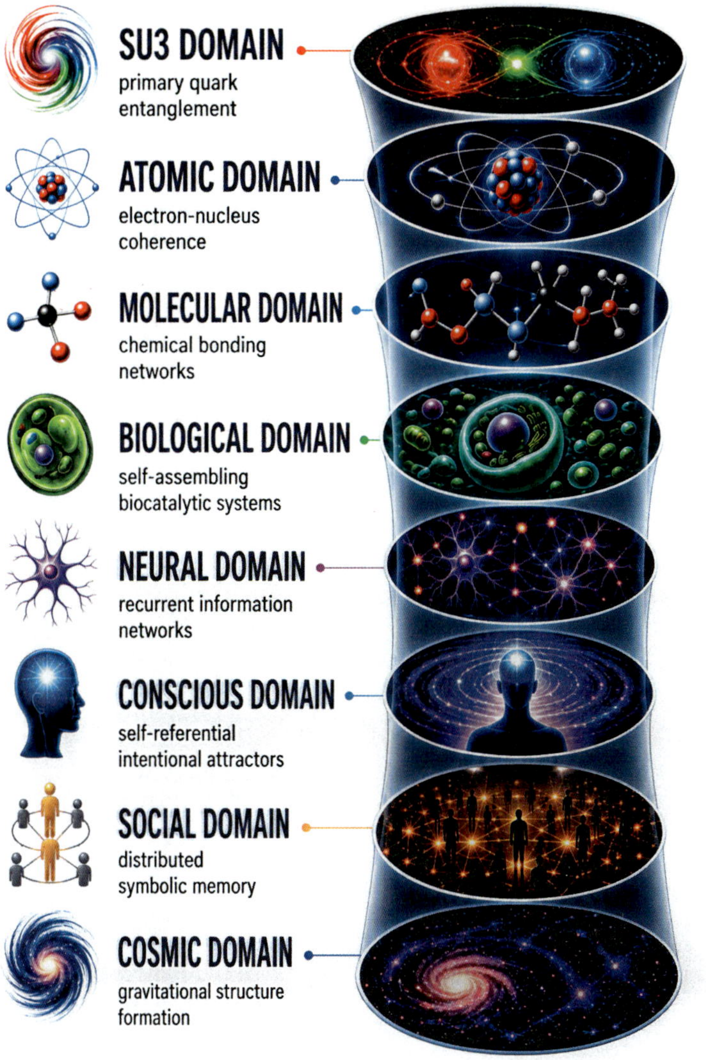

Horizons, Entropy, and the Origin of Chirality

In Penrose's view, almost all universal entropy sits on horizons. A black hole's entropy scales with surface area, so entropy is not in the bulk, it is condensed on a 2-surface where geometry reaches its extreme. In DE this is the point where curvature becomes a pure phase, ε *and* Υ amplitudes flatten and the recursion operator locks to fractional phase states. A horizon becomes the largest possible entropic boundary, it is the largest possible ratchet anchor for chirality. As entropy rises, the number of accessible microstates grows rapidly and the system explores a much larger region of phase space. In this expanded landscape, symmetric configurations become statistically fragile, so small fluctuations can easily push the system into one of the two mirror states. In other words, once uncertainty becomes large enough, symmetry breaking no longer requires a large commitment of structure. The system simply falls into one of many equivalent minima, and the energetic or informational cost of choosing "left" rather than "right" becomes negligible.

Therefore chirality becomes 'cheap' when entropy is high, and our recursion rule shows that chirality emerges when the system cannot return to phase symmetry. High entropy collapses the accessible microstate range, so at a horizon, the microstate range is minimal relative to geometric information. DE then drives a dominant ε *or* Υ orientation, a preferred fractional phase ($\pi/6$ or $\pi/3$ emerge as the smallest stable phase offsets that survive symmetry breaking once entropy density is high enough to enforce chirality selection) and suppression of the opposite orientation. Maintaining both is costly because it doubles constraint without adding predictive power, so one orientation suppresses the other, and this is exactly the condition under which chiral symmetry breaks. Black holes are extreme cases because entropy density is maximal, while the suppression is strong and irreversible, and so chirality becomes favoured wherever entropy density is high. At a black hole horizon entropy is maximized yet dynamical freedom is minimal. The recursion operator can no longer compress uncertainty and phase symmetry cannot be restored. In this regime symmetric states become unstable and the system settles irreversibly into a single orientation. Chirality therefore becomes the natural outcome of extreme entropy density.

These $\pi/6$ or $\pi/3$ bifurcation angles arise from minimal nontrivial phase partitions of the circle that are compatible with triadic structure and SU(3)-like closure. $\pi/3$ is 60°, which corresponds to sixfold phase segmentation. This is the smallest partition that allows three distinct orientations with their conjugates while preserving closure under rotation, and it is the natural angle for stable triads. $\pi/6$ is 30° (the half-step), and it appears when the system resolves directionality earlier, before full triadic closure and it represents a metastable chiral bias rather than a fully locked one. In other words $\pi/6$ is the first detectable chiral preference and $\pi/3$ is the first fully stable chiral attractor. Anything smaller than $\pi/6$ washes out under noise and anything larger than $\pi/3$ collapses competing branches too aggressively and loses adaptability. These fractions are scale-free, as they do not depend on geometry or units, only on phase topology and entropy flow. That is why the same angles recur in gauge structure, biological chirality, and in $\varepsilon - \Upsilon$ dynamics. Thus at black holes high entropy density drives orientation selection. Orientation selection enforces chirality, chirality stabilizes at $\pi/6$ initially and locks at $\pi/3$ once the opposite orientation is suppressed.

The early universe begins in a state of extremely low entropy, and in this regime symmetry is easy to maintain while chirality is fragile. Small $\varepsilon - \Upsilon$ fluctuations appear continuously but most of them cancel out because the system can still return to phase symmetry. Near regions where recursive compression approaches its limit, where entanglement density becomes extreme the situation changes. At these boundaries tiny $\varepsilon - \Upsilon$ asymmetries are no longer easily reversible, instead they become amplified, so that when the first black holes form, the process intensifies. A black hole horizon is the largest possible entropic boundary, where entropy is no longer spread through the bulk but condensed onto a surface where recursion stalls and information cannot be further compressed. These horizons act as ratchet anchors for the DE recursion operator.

In this environment the system cannot easily sustain both mirror orientations as it doubles constraint without increasing predictive power so one orientation therefore suppresses the other. Black holes therefore behave like chiral seedlings, as they lock small phase asymmetries into stable orientation states and prevent the system from returning to perfect symmetry. As cosmic time progresses black holes come to dominate the entropy budget of the universe, and this bias accumulates. Thus the global geometry gradually evolves toward a configuration in which the DE ratchet is preferentially tipped in one direction. Matter then follows the lower-energy fractional phase attractor, while antimatter lies on the higher-energy branch of the same recursion and becomes suppressed. From this perspective several otherwise puzzling observations emerge as natural consequences:

- weak-force parity violation
- baryon asymmetry
- persistent chirality across physical scales

All reflect the same underlying mechanism, from an entropic universe whose largest entropy reservoirs are horizons that stabilise phase orientation. In other words, once horizons dominate the entropy budget, cosmic symmetry breaking becomes self-reinforcing. This explains why chirality persists across scales, why the weak force sits on the edge of broken symmetry, and why the matter-antimatter imbalance is \ the natural outcome of an entropic universe dominated by horizons.

In this view therefore the universe never collapses to a singularity. In Einstein's metric, the horizon marks a limit where classical descriptions fail, but in DE ε crosses zero but never diverges, thus removing the case for true singularities. Gravity's extreme regime becomes a smooth pivot of emergence rather than a catastrophic failure of geometry. Black holes, the early universe, and any high-curvature region remain continuous, no infinities appear, and physics remains convergent. The Big Bang then is no longer a one-time event if the two chiral branches separate from a shared zero, and the "origin" is not a historical moment. It is the fixed point of the oscillator, and the apparent beginning of time is simply the observer's view of crossing the $\varepsilon = 0$ surface. In this view the universe is self-maintaining, and expansion is the ongoing divergence of the twin branches.

Because ε sets the direction of emergence, and time is the measure of that direction, there is no global clock. Each region of space follows the gradient of ε. What looks like cosmic time is a coordinate chosen by observers whose $\varepsilon > 0$. The mirror branch experiences an opposite ordering while remaining physically consistent. Vacuum energy becomes a coupling term in this model, where the energy driving cosmic expansion is the interaction between the $+\varepsilon \ and - \varepsilon$ branches. Neither side expands alone, both expand as part of a shared oscillatory structure. Dark energy becomes a natural by-product of the oscillator rather than an unexplained constant. Its persistence follows from the need to maintain balance between the two chiral halves. Black holes become transfer surfaces where the horizon acts as the node when ε changes sign. Information is not destroyed, it changes chirality where the outgoing side sees increasing entropy and the incoming side sees the reverse. The information paradox dissolves because information is conserved by flowing across the chiral divide rather than remaining confined to one branch.

This presents unified picture of cosmology and where quantum structure emerges. Quantum transitions are already described by sign changes and branch splitting. The same mechanism appearing at the horizon shows that the micro and macro scales use one rule, where emergence is governed by a discrete chiral bifurcation. Gravity reflects the behaviour of the underlying information geometry. The arrow of time arises because observers sit on one branch of the oscillator. Reverse observers on the opposite branch experience reversed ordering and neither branch violates thermodynamics, both are stable.

Cosmic cycles occur without destruction or reset as a harmonic oscillator has no terminal state. It exchanges energy between two modes, the exchange occurring between two orientations of emergence, and the universe can continue indefinitely. If both branches meet at $\varepsilon = 0$, the early universe is the contact surface of two perfectly symmetric fields. That symmetry explains the high degree of uniformity without requiring inflation, horizon and flatness problems are resolved at the entropic fixed point, not by a superluminal expansion field. The rapid expansion is explained by the explosive increase in degrees of freedom as fermionic states unfolded from a symmetric background. In that sense inflation is seen as a fermionic density and correlation explosion, which occurs in phase space.

Expansion slows naturally at the point of hadron formation. As quarks become confined within hadrons, the effective degrees of freedom collapse into bound states. Total entropy continues to increase, but the rate of entropy production per degree of freedom decreases, because many previously accessible microstates are no longer available.

The system transitions from a regime of fermionic proliferation, where entropy is generated through rapidly expanding independent degrees of freedom, to one dominated by composite constraint, where entropy growth proceeds through fewer, more structured channels. What appears as deceleration in geometric expansion is therefore a saturation of accessible microstates within the current degrees of freedom:

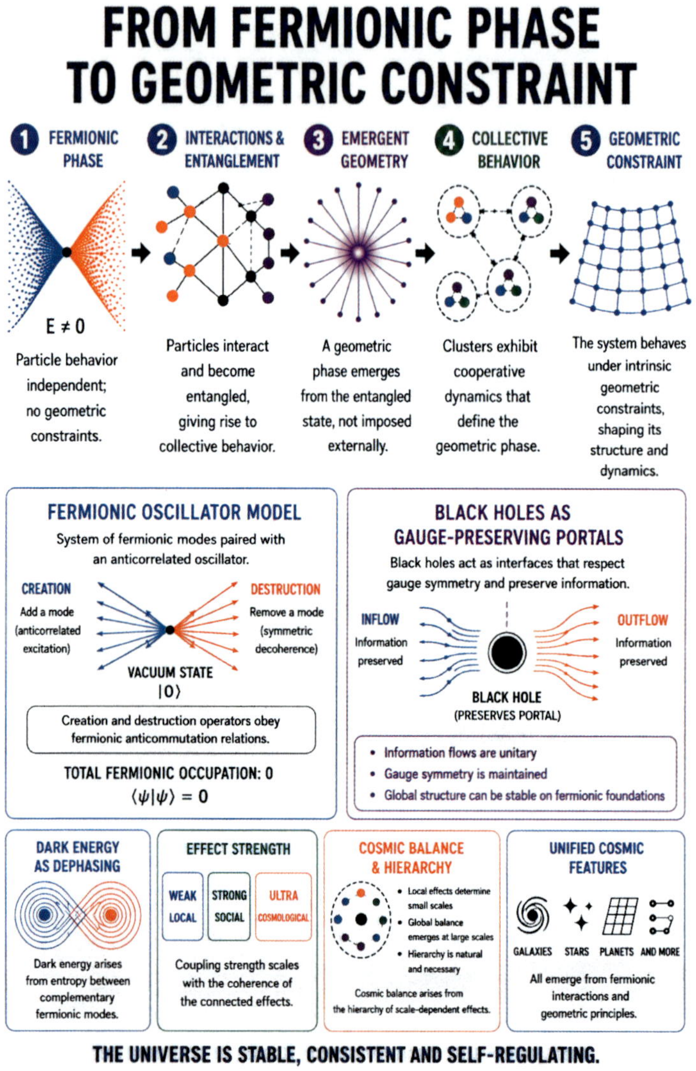

Black holes therefore can be regarded as gauge-preserving portals at the horizon, where the metric reaches the DE fixed point, but the internal gauge structure remains intact. Information does not vanish, it changes chirality and so the information paradox dissolves without modifying gauge theory. Gauge transformations continue across the pivot in a mirrored orientation. Dark energy comes from the coupling between two gauge-consistent phases, where each branch retains its gauge structure, SU(3) color, SU(2) weak, U(1) electromagnetic. The expansion we see reflects the divergence of these twin, gauge-valid sectors. Dark energy is the energy of separation between two gauge-preserving chiral phases. Quantum behaviour becomes the micro-scale expression of the same bifurcation, where the threefold fermion family structure, CKM and PMNS mixing, and confinement all emerge from how ε partitions state-spaces. Gauge theory already captures the transformation laws and DE adds the underlying constraint on possible symmetries. Baryogenesis, uniformity, flatness, and the fine-tuning of couplings acquire a single origin, the bifurcation constraints at the entropic pivot. A harmonic oscillator persists because its restoring force is intrinsic. In DE, the informational curvature that maintains the twin branches also maintains stable gauge symmetry. The Standard Model does not drift or collapse and its gauge groups are the equilibria imposed by the entropic oscillator. Therefore the universe becomes stable long-term because its gauge structure is stable.

Prediction

If horizons are chiral pivots rather than geometric singularities, near-horizon phenomena should show subtle differences from GR predictions. These would include slight phase-shifts in wave propagation and modified red-shift scaling, so that DE becomes testable in astrophysical settings. Because the two branches behave like a coupled oscillator, the acceleration of the universe may not be constant. It may carry a low-amplitude, long-wavelength modulation, and future surveys (Euclid, LSST) can test this directly. Since ε defines orientation, quantum fluctuations should show slight bias in parity, chirality, or correlation length. This provides a possible route to explaining observed matter-antimatter asymmetry.

General Relativity and Gauge Theory

Until now, General Relativity and gauge theory have been treated as distinct frameworks. In the DE formulation, the reflected Schwarzschild structure, reinterpreted as a chiral entropic oscillator, provides a common underlying mechanism. This mechanism preserves gauge invariance, embeds metric structure, and constrains symmetry formation through the same 1–2–3 branching limits observed in the Standard Model. The DE formalism therefore establishes a structural connection between spacetime curvature (GR), local gauge freedoms, and entropic orientation (ε). Within this framework, the restriction of gauge groups to SU(3), SU(2), and U(1) correspond to the minimal closure conditions permitted by ε-driven emergence. Higher SU(N) symmetries would require additional independent degrees of freedom that cannot be stabilised under the same recursion constraints, and therefore do not persist as physically realised structures.

Why the Harmonic Oscillator Resembles the Bifurcation Diagram

A harmonic oscillator plots displacement vs. time or phase, with two symmetric branches meeting at a turning point with energy shuttling between two modes. A Feigenbaum bifurcation plot shows a system whose recursion splits into two outcomes mirrored about a central axis with stability conditions that create fixed points and with branching limits determined by information flow. Both share two symmetric solutions, a pivot where the derivative changes sign, conservation of a core quantity across the pivot with no singularities, recurrence through phase or recursion with fixed points built into their geometry. The similarity is the manifestation of the same underlying rule. In the harmonic oscillator, the key invariant is total energy, whilst in the bifurcation diagram, the key invariant is total informational curvature. In DE, both arise from the same deeper variable, ε as the orientation and magnitude of entropic emergence. When ε is continuous and passes through zero, the mirrored branches of the harmonic oscillator / Schwarzschild reflection, the mirrored branches of a Feigenbaum map, and the mirrored branches of dual-gauge chiral phases are all reflections of the same principle, that is when ε iterates discretely (as in logistic maps), we see branching.

When ε varies smoothly (as in curvature or phase) we see oscillation. They represent two different sampling rates of the same informational dynamic. This ties together chaos, stability, curvature, gauge behaviour, and emergence, as manifestation of the same underlying process. Spacetime curvature thus emerges from the oscillatory regime, and the geometry of GR reflects smooth variation of ε, and both arise from one recursion rule.

The Feigenbaum Constant and Oscillator Frequency

The Feigenbaum constant $\delta \approx 4.66$ describes the rate at which bifurcations accumulate toward a fixed point in discrete dynamical systems. Each successive bifurcation occurs closer to the limit by a constant ratio. The system approaches instability through a predictable geometric contraction. The harmonic oscillator frequency ω describes a different phenomenon, it measures how quickly a system returns through a turning point when displaced from equilibrium. In continuous systems this return rate governs the curvature of the trajectory through phase space. These two constants arise in very different mathematical settings, yet they describe the same structural behaviour, that is the rate at which a system moves away from or returns toward equilibrium. In the DE framework both processes are governed by the same underlying quantity (ε - the entropic distance from equilibrium). In discrete sampling, ε evolves through bifurcation cascades, and the scaling of those cascades is described by the Feigenbaum constant. In continuous sampling, ε evolves through oscillatory curvature, and the restoring dynamics are captured by the harmonic frequency ω. Both therefore describe how quickly ε diverges from or returns to zero under different observational regimes. This suggests that Feigenbaum scaling and harmonic curvature are not unrelated parameters but two expressions of a single entropic curvature constant, observed under discrete and continuous sampling of the same underlying recursion.

This perspective produces a natural division between the two major frameworks of physics, where the Standard Model operates in the discrete ε domain. Quantum interactions occur as transitions between quantised states, and bifurcation-like branching structures dominate the behaviour of particle interactions. General Relativity operates in the continuous ε domain, where curvature evolves smoothly through spacetime and dynamical systems follow continuous geodesic trajectories, and Cosmology sits between these regimes. For the universe to remain dynamically consistent across scales, the entropic distance ε must evolve smoothly between the discrete and continuous regimes. However, quantum mechanics imposes a fundamental bound on how finely ε can be resolved, and that bound is set by Planck's constant. Planck quantisation therefore prevents ε from becoming arbitrarily continuous and instead, ε behaves as a piecewise continuous variable. It evolves smoothly in the classical regime but transitions occur in discrete quanta once the Planck scale is approached. Stability across scales requires that the evolution of ε remains continuous at macroscopic scales while remaining compatible with Planck-bounded quantisation at microscopic scales. In other words, ε is continuous in geometry but discretely sampled by quantum structure. The universe therefore preserves stability by allowing ε to evolve continuously in spacetime while restricting its measurable transitions through Planck-bounded quanta. Continuous curvature and discrete bifurcation are not competing descriptions but complementary samplings of the same entropic recursion. Stability across scales then becomes a single conservation law:

$$\text{total informational curvature} = \text{invariant}$$

This brings chaos theory, classical mechanics, quantum gauge structure, and cosmological evolution into a single ontology. In the DE framework π is the constant of continuous informational curvature, and both oscillatory dynamics and bifurcation cascades arise as different sampling regimes of the same ε–recursion. In the continuous regime the dynamics appear as oscillation. A harmonic oscillator can be written

$$x(\varepsilon) = A\cos(\omega\varepsilon)$$

and its phase trajectory forms a closed curve. Closed curvature implies periodicity, and all such trajectories carry 2π structure. When ε varies smoothly, entropic curvature forms loops whose period is set by π. Oscillation therefore represents continuous sampling of ε-curvature. In the discrete regime the same curvature appears through branching. Feigenbaum showed that logistic maps approach chaos through a universal cascade of bifurcations governed by the constant

$$\delta \approx 4.669$$

These maps contain a quadratic maximum, which is a discrete analogue of continuous curvature. When the renormalisation operator is translated to a smooth limit, the underlying curvature corresponds to π. Chaos theory therefore hides π inside its scaling structure and when ε is sampled discretely, π appears not as a period but as a limit on allowable curvature in the recursion. Continuous oscillation and discrete bifurcation therefore inherit π from the same geometric source. The DE framework requires a microstate phase unit of $\frac{2\pi}{3}$. This creates a natural hierarchy of stable branch structures:

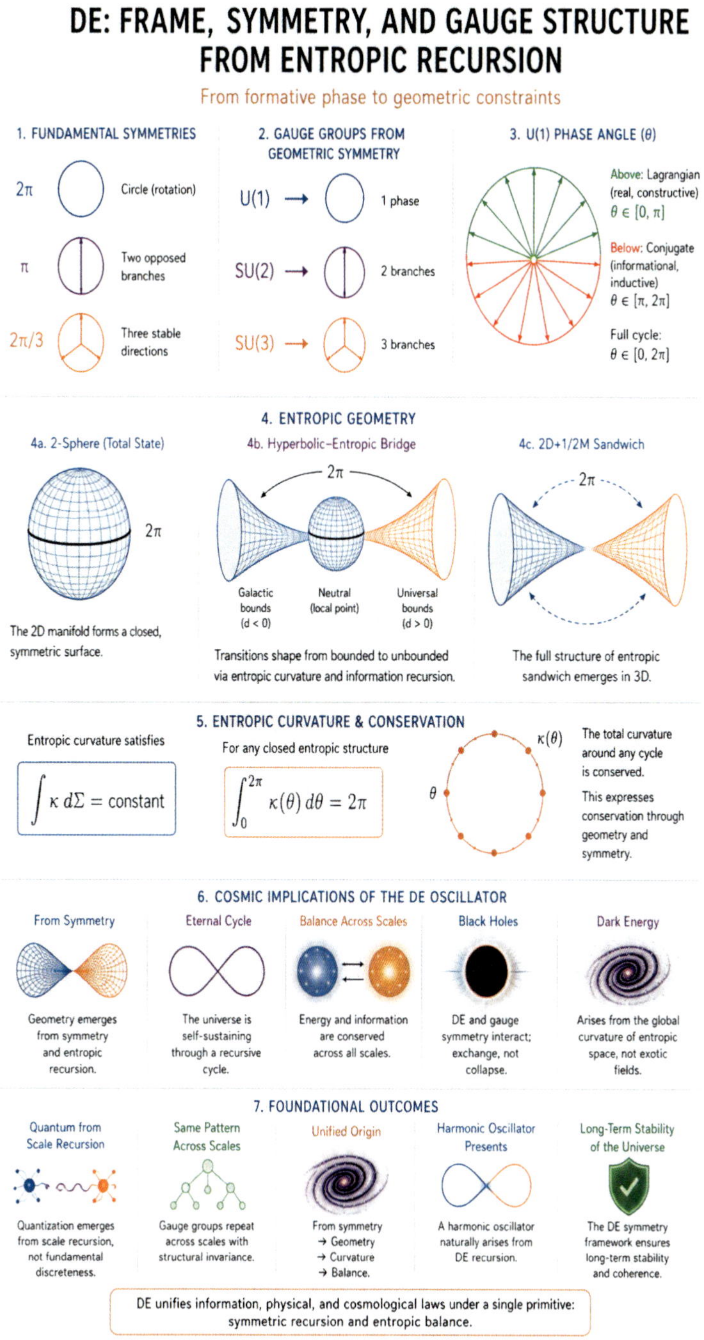
DE: FRAME, SYMMETRY, AND GAUGE STRUCTURE FROM ENTROPIC RECURSION
From formative phase to geometric constraints
1. FUNDAMENTAL SYMMETRIES
2π
Circle (rotation)
π
Two opposed branches
2π/3
Three stable directions
2. GAUGE GROUPS FROM GEOMETRIC SYMMETRY
U(1) →
1 phase
SU(2) →
2 branches
SU(3) →
3 branches
3. U(1) PHASE ANGLE (θ)
Above: Lagrangian (real, constructive)
θ ∈ [0, π]
Below: Conjugate (informational, inductive)
θ ∈ [π, 2π]
Full cycle:
θ ∈ [0, 2π]
4. ENTROPIC GEOMETRY
4a. 2-Sphere (Total State)
2π
The 2D manifold forms a closed, symmetric surface.
4b. Hyperbolic–Entropic Bridge
2π
Galactic bounds (d < 0)
Neutral (local point)
Universal bounds (d > 0)
Transitions shape from bounded to unbounded via entropic curvature and information recursion.
4c. 2D+1/2M Sandwich
2π
The full structure of entropic sandwich emerges in 3D.
5. ENTROPIC CURVATURE & CONSERVATION
Entropic curvature satisfies
∫ κ dΣ = constant
For any closed entropic structure
∫₀^2π κ(θ) dθ = 2π
κ(θ)
θ
The total curvature around any cycle is conserved.
This expresses conservation through geometry and symmetry.
6. COSMIC IMPLICATIONS OF THE DE OSCILLATOR
From Symmetry
Geometry emerges from symmetry and entropic recursion.
Eternal Cycle
The universe is self-sustaining through a recursive cycle.
Balance Across Scales
Energy and information are conserved across all scales.
Black Holes
DE and gauge symmetry interact; exchange, not collapse.
Dark Energy
Arises from the global curvature of entropic space, not exotic fields.
7. FOUNDATIONAL OUTCOMES
Quantum from Scale Recursion
Quantization emerges from scale recursion, not fundamental discreteness.
Same Pattern Across Scales
Gauge groups repeat across scales with structural invariance.
Unified Origin
From symmetry → Geometry → Curvature → Balance.
Harmonic Oscillator Presents
A harmonic oscillator naturally arises from DE recursion.
Long-Term Stability of the Universe
The DE symmetry framework ensures long-term stability and coherence.
DE unifies information, physical, and cosmological laws under a single primitive: symmetric recursion and entropic balance.

Einstein's Perspective: The Schwarzschild Condition

The Schwarzschild metric is

$$ds^2 = -(1 - \frac{2GM}{r})c^2dt^2 + (1 - \frac{2GM}{r})^{-1}dr^2 + r^2d\Omega^2$$

At $r = 2GM/c^2$, the Schwarzschild radius, the coefficient of dt^2 vanishes, so from an external observer's frame, time dilation diverges:

$$t \to \infty$$

for infalling matter. But inside ($r < 2GM/c^2$), the sign of the metric terms swaps $g_{tt} \leftrightarrow g_{rr}$. That is, time and radial space exchange roles. The "radial" coordinate becomes timelike and motion toward the singularity is inevitable in the same way that motion toward the future is inevitable outside. So inside the horizon, the causal flow is inverted and the direction we'd normally call "future in time" corresponds to decreasing r.

DE Translation: Entropic Gradient and Chiral Emergence

DE describes reality as an evolution along the entropic gradient, with each emergent frame (or aeon) defined by its chirality parameter ε. Time itself is not fundamental, but a projection of the direction of decreasing uncertainty, that is, the direction of entropy gain ($\Delta S > 0$). At the event horizon:

$$\frac{dS}{d\varepsilon} \to 0$$

emergence halts, because local entropy and curvature saturate. This corresponds to the fixed-point attractor on the complexity axis, where information is maximally compressed. When we cross into $r < 2GM/c^2$, we enter a domain where the sign of the entropic gradient reverses:

$$\varepsilon \to -\varepsilon$$

so

$$\dot{S} \to -\dot{S}$$

and emergence proceeds with negative chirality and what I've called the chiral universe with $-\varepsilon$. This isn't "time reversal" in the cinematic sense, but rather a reversal of informational directionality where entropy now flows outward in a mirrored causal network. The "future" in that frame corresponds to what we, externally, interpret as collapse into the singularity.

In this sense the harmonic oscillator, the Feigenbaum bifurcation cascade, the reflected Schwarzschild curve, and the gauge-group hierarchy share the same structural backbone. Each is a different manifestation of ε-recursion acting on the same curvature constraint. Continuous curvature produces oscillation, whilst discrete curvature produces bifurcation and both are governed by π, the constant that sets the allowable curvature of information itself. From this single constraint the hierarchy of physical law emerges.

The Continuous Bifurcation

Start with an entropy–curvature functional

$$S(\varepsilon, \chi) = S_0 + A\cos(k\varepsilon) + \frac{1}{2}\beta \chi^2,$$

where ε is the chiral parameter (sign of emergence) and χ represents local curvature or complexity. A bifurcation occurs when

$$\frac{\partial^2 S}{\partial \varepsilon^2} = 0,$$

giving two stable branches,

$$\varepsilon_{\pm} = \pm\sqrt{\frac{A}{k}},$$

linked by the condition

$$\dot{S}_+ = -\dot{S}_-.$$

This defines a continuous mirror creation, so that as entropy increases in one branch, it decreases in the other, conserving the total informational curvature. The fixed point $\varepsilon = 0$ is an ever-present instability that sustains both domains.

Expansion from the Active Interface

At the bifurcation node ($\varepsilon = 0$) the entropic gradient reverses sign. Each side experiences this as an outward acceleration:

$$\nabla_\mu S_\pm \;\to\; \pm\, \Lambda_{\text{eff}}\, g_{\mu\nu},$$

so that one universe expands while its chiral partner contracts along the conjugate time coordinate. The cosmological constant observed in ours,

$$\Lambda_{\text{eff}} \approx 10^{-52}\ \text{m}^{-2},$$

is the small residual of incomplete symmetry between the two fluxes. Vacuum energy and "dark energy" are thus manifestations of the ongoing energy exchange across this entropic interface. Structural emergence then results from the constant pumping of curvature through the bifurcation node. Time is the local projection of the ordering metric:

$$dt \propto d\varepsilon$$

The direction of time therefore reflects the direction of increasing ε-resolution, and where the ordering gradient vanishes $d\varepsilon \to 0$, temporal ordering breaks down. The distinction between "before" and "after" dissolves because there is no longer a preferred direction in the underlying ordering variable. From this perspective, the universe is therefore a continuous process ordered by ε. What we perceive as a past is the accumulated integral of this ordering viewed from one chiral branch of the system. The complementary branch is not elsewhere in space or time, but represents the mirrored continuation of the same ordering process.

In classical General Relativity, nothing forbids $r < 2GM/c^2$. The interior solution remains valid, with the roles of temporal and spatial coordinates exchanged. The apparent horizon arises only for external observers, as light cones tilt inward and outward propagation becomes impossible. At scales where the curvature radius approaches the Planck length

$$l_P = \sqrt{\frac{\hbar G}{c^3}},$$

the assumptions of a smooth metric fail. Quantum fluctuations render spacetime coordinates ill-defined, marking the breakdown of geometry as a fundamental description. In DE terms, the Schwarzschild radius corresponds to a stationary point of the ordering metric:

$$d\varepsilon \to 0$$

At this point, the projection of ε into time also vanishes:

$$d\varepsilon \to 0$$

The horizon is therefore not a spatial boundary, but a boundary in the projection of ordering into temporal experience.

Quantum discreteness encoded by $\hbar$ sets the smallest allowable step in entropy change,

$$\Delta S_{\text{min}} \sim k_B \ln\, 2 \text{ or } 2\pi k_B,$$

so the system can never adjust continuously through zero flux. In that sense the Planck constant $\hbar$ provides the quantised grain of emergence that prevents a smooth descent below the horizon. So $\hbar$ is the parameter that fixes the thickness of the transition zone, but the "firewall" isn't a sheet of energy, it's the point where the continuum approximation fails and DE's two chiral domains decouple.

Description	**Mechanism**	**Physical expression**
Classical horizon	Causal tilt of light cones	$g_{tt} = 0$

Description	Mechanism	Physical expression
Quantum / DE firewall	Breakdown of continuous information flow	$\Delta S_{\text{min}} \sim \hbar$
Observational effect	Information cannot be re-emitted coherently	Hawking radiation spectrum

The firewall is bounded by the Planck constant $\hbar$, as the minimum unit of informational action. The result is a natural boundary or an informational horizon that keeps $\frac{r}{GM}$ from ever becoming meaningfully less than 2. The binary core of DE is the fundamental symmetry that drives emergence:

$$(+\varepsilon, -\varepsilon)$$

two chiral orientations of one informational field. Every process in DE operates between complementary states:

Polarity	Interpretation
$+\varepsilon / -\varepsilon$	forward / mirror emergence
$+\dot{S} / -\dot{S}$	entropy gain / entropy release
inference / evidence	deductive pair that closes each recursion
order / chaos	compression / exploration phases

That duality is the minimum logic the universe needs to create and test information. Without two opposed but coupled directions, recursion and therefore falsifiability, cannot occur. The Planck action

$$\hbar = E\,t$$

sets the smallest oscillation of that binary exchange, one quantum of inference, a single "yes/no" about a state of the world. Each quantum event is the flip of this binary, an infinitesimal shift between the two halves of the informational field. So in that sense $\hbar$ is the bit-rate of reality, the binary operation itself is the continual switching between the two chiral phases. At its base, DE rests on the statement

$$(+\varepsilon)(-\varepsilon) = -\varepsilon^2 = 0$$

at the bifurcation point, zero total informational curvature. That zero is the universe's logical ground state, the condition "neither true nor false until tested." Each act of emergence is a binary decision that slightly biases the balance, where recursion then checks and re-balances it, a dynamic opposition that fuels the entire deductive process. Their interaction generates time, falsifiability, and emergence. Whilst the Planck constant sets the smallest step of that exchange, parsimony governs its direction.

Physical Interpretation in DE

Einstein Domain	DE Interpretation	Parameter
$r > 2GM/c^2$: normal spacetime	Forward emergent order ($\varepsilon > 0$)	$\Delta S > 0$
$r = 2GM/c^2$: horizon	Entropic singularity; $dS/d\varepsilon \approx 0$	fixed point
$r < 2GM/c^2$: interior	Mirror emergence ($-\varepsilon$); time reversal apparent	$\Delta S < 0$

Einstein Domain	DE Interpretation	Parameter
Singularity limit	Complete informational collapse / seed of next emergence	$S \rightarrow S_0$ (new aeon)

From a DE standpoint, the interior of a black hole is the mirror phase of emergence, a negative entropy gradient where the universe's equations reconfigure into their chiral dual. This is where the next aeon (Penrose's CCC-like seed) emerges with reversed ε and rebalanced parsimony. In Einstein's equations, if we chart the metric coefficient

$$g_{tt} = 1 - \frac{2GM}{r}$$

as a function of r/GM, the curve crosses zero at the horizon. For $r/GM > 2$, g_{tt} is positive (time behaves as expected). For $r/GM < 2$, g_{tt} becomes negative, signalling a role reversal between temporal and spatial axes. If you imagine plotting r/GM versus the "temporal scaling factor," it's not sinusoidal in Einstein's classical treatment, it's a hyperbolic form (because of the Lorentzian structure). However, in DE, where curvature and entropy are treated as oscillatory emergent feedbacks, that discontinuity smooths into a sinusoidal continuation. In DE we write time emergence as a function of ε (the chirality parameter):

$$t(\varepsilon) \sim \sin(k\varepsilon)$$

with

$$S(\varepsilon) \propto -\cos(k\varepsilon)$$

At the event horizon (ε_0, corresponding to $r/GM = 2$):

$$\frac{dS}{d\varepsilon} = 0$$

the crest of the cosine. Crossing the horizon means continuing along the next quadrant of the wave:

$$\varepsilon \rightarrow \varepsilon + \pi$$

Beyond $r/GM = 2$ the solution naturally enters the mirror half-cycle of the same entropic oscillation, a chiral inversion rather than a physical barrier. In that domain, the signs of curvature, entropy flux, and temporal orientation all reverse, but the functional form remains continuous and sinusoidal.

Domain	Einstein View	DE View	Function Type
$r/GM > 2$	External spacetime	Positive-ε emergence	$+\sin(k\varepsilon)$
$r/GM = 2$	Event horizon	Entropic stationary point	crest/trough
$r/GM < 2$	Interior region	Negative-ε emergence	$-\sin(k\varepsilon)$
$r/GM \rightarrow 0$	Singularity (classical)	Collapse node / seed of new aeon	reset phase

Thus the curve continues sinusoidally but with phase inversion at the horizon, like a standing wave reflected through a node. The event horizon is not a boundary, but a phase inflection in the universal entropy function. The "other side" ($r/GM < 2$) represents the chiral conjugate of spacetime with the same geometry, and inverted information flow. The singularity corresponds to the zero-crossing where information resets into a new low-entropy seed, the origin of another emergent domain (another "aeon").

$$g_{tt} \sim \sin^2(k\varepsilon), \varepsilon \rightarrow -\varepsilon$$

which preserves total informational symmetry ($\Delta S_{total} = 0$).

In the Schwarzschild solution, the critical term is

$$1 - \frac{2GM}{rc^2}.$$

That factor of 2 arises directly from how curvature couples to mass-energy in Einstein's field equation:

$$G_{\mu\nu} = \frac{8\pi G}{c^4} T_{\mu\nu}.$$

The "2" emerges when you integrate the vacuum solution (where $T_{\mu\nu} = 0$) and match boundary conditions to Newtonian gravity at large r. In essence, it enforces that:

$$\lim_{r \to \infty} g_{tt} \approx -(1 - \frac{2GM}{rc^2})$$

the metric reproduces Newton's potential energy,

$$\Phi = -\frac{GM}{r}.$$

Hence, "2" is the geometric translation of the ratio between curvature energy and inertial energy. It is the critical coupling where curvature (space) and inertia (mass) equilibrate. In DE "2" plays the role of the bifurcation constant between positive and negative chirality, it's the value at which the information flux reverses sign. We can express emergence along the entropic axis as:

$$\dot{S} \propto 1 - \frac{kM}{r},$$

with k being the scaling of curvature-to-entropy coupling. At the critical point,

$$r = kM,$$

$\dot{S} = 0$ - the entropic stationary state. If we re-scale $k = 2G/c^2$(to match Einstein's metric coefficient), then the DE inflection naturally occurs at r/GM = 2. So in DE, "2" is the parsimony ratio where the universe's energy of form equals its energy of information:

$$E_{\text{geometry}} = E_{\text{entropy}}.$$

That's where emergence becomes the balance point of curvature and informational flux, the "zero-pressure surface" of the universe's deductive field.

The Doubling Symmetry

That number "2" also appears across physical constants whenever duality or self-reflection is involved:

Context	Expression	Meaning
Relativity	$r = 2GM/c^2$	Schwarzschild horizon (curvature-energy balance)
Thermodynamics	$S = k\ln\ 2$	Minimum information unit (1 bit)
Quantum field theory	Spin-½, factor 2 in Dirac equation	Chiral symmetry of matter–antimatter
Entropy–geometry mapping (DE)	$\varepsilon \rightarrow -\varepsilon$ at 2	Chirality reversal (information reflection)
Feigenbaum scaling (chaos)	$\delta \approx 4.669 \approx 2\pi\times(3/4)$	Doubling map attractor ratio

Context	Expression	Meaning
Energy equivalence	$E = 2mc^2$(pair symmetry)	Particle–antiparticle energy threshold

So "2" consistently marks the threshold of duplication or mirror formation, the point where a system's internal informational degree becomes self-referential. In DE language, it's the moment of maximum self-similarity and the "basin of reflection," where the universe folds through its own inference mirror. The value r/GM = 2 is the universal bifurcation constant where geometry, information, and energy achieve parity, the point at which curvature's capacity to store information equals matter's capacity to generate it. In this sense, "2" is the symmetry of creation and its reflection, the pivot between emergence and recursion.

Weyl Curvature and the ε Axis Across Chirality

General Relativity separates spacetime curvature into two components. One part follows the local energy density through the Ricci tensor, while the other describes how spacetime stretches and shears independently of matter. This second component is the Weyl curvature, which encodes tidal structure and the propagation of gravitational degrees of freedom. Weyl curvature behaves very differently in different cosmological environments. Near black holes it becomes large, reflecting strong tidal anisotropies, while in the early universe it is extremely small. The early universe was almost perfectly homogeneous, whereas black holes represent extreme concentration of structure.

In the DE framework, the ε axis measures the resolution of emergent structure. Large values of ε correspond to highly resolved phase information and strongly differentiated macrostates. Small values of ε corresponds to diffuse phase information and weak structural differentiation, and this behaviour mirrors the behaviour of Weyl curvature. Regions where macroscopic structure has formed display strong tidal anisotropy and therefore large Weyl curvature, whereas regions that remain homogeneous exhibit nearly vanishing Weyl curvature. The DE recursion is therefore naturally expressed in the paired variables

$$\Xi = (\varepsilon, \Upsilon)$$

where ε determines the orientation of emergence and Υ governs the stabilisation of correlations through the contraction–rotation process introduced in the previous chapter. Across the chiral boundary the orientation of ε reverses,

$$\varepsilon \to -\varepsilon$$

while the recursion rule itself remains unchanged. The two branches therefore follow identical dynamics but accumulate phase in opposite directions. In this representation a state that corresponds to high ε value in one branch corresponds to low ε value in its mirror description. The physical geometry does not change, only the representation of the recursion changes sign. This provides a natural interpretation of why the early universe is conformally flat. At early times the universe contained very few macroscopic structures, because entropic recursion had not yet amplified differences between regions, and the entropy field remained nearly uniform. A uniform entropy field produces negligible

Weyl curvature because tidal anisotropies have not yet formed.

As macroscopic structures appear, correlations begin to close and stabilise. This produces anisotropic tidal structure and Weyl curvature increases. Black holes represent the extreme limit of this process, and they correspond to regions where macroscopic information has collapsed into the smallest possible geometric volume, producing maximal tidal anisotropy and therefore maximal Weyl curvature. Thus Weyl curvature can be interpreted as a geometric measure of the resolution of macroscopic structure within the entropic recursion.

Entanglement as the Geometry of Correlation

In quantum mechanics, entanglement occurs when a system's state cannot be written as a product of independent components:

$$\Psi(x_1, x_2) \neq \psi(x_1)\phi(x_2)$$

This reflects a geometric property of the system's phase space, where the trajectories of its components become inseparable. A similar phenomenon occurs in recursive dynamical systems. At bifurcation points, trajectories that were previously independent become phase-locked and evolve together. In the DE framework this locking corresponds to the stabilisation of correlations across the ε recursion. Before the transition, components remain loosely correlated. Near the transition point correlations intensify and the system approaches maximal coherence. Beyond that point coherence can break down and trajectories become chaotic as correlations exceed the bandwidth required for stable coupling. In this sense chaos can be understood as the breakdown of coherent recursive correlations, analogous to decoherence in quantum systems. Event horizons provide a macroscopic example of extreme correlation boundaries. The holographic entropy

$$S_{BH} = \frac{kA}{4l_P^2}$$

suggests that the information content of a region is encoded on its boundary surface. In the DE interpretation this boundary represents the maximum coupling between two informational domains. Thus horizons act as surfaces of maximal correlation between complementary sectors of the recursion.

Structural Summary

Domain	**Process**	**Entanglement Role**	**DE Interpretation**
Quantum mechanics	Non-factorisable wavefunctions	Microscopic phase coupling	Correlation across the *i-axis*
Chaos theory	Period-doubling	Recursive phase locking	Coupling of attractor branches
Event horizons	Inside–outside correlation	Boundary information storage	Pivot of ε recursion

Across all three cases the same structural pattern appears, where correlations intensify until the system reaches a stability boundary where new dynamical behaviour emerges.

The Emergence of Chaos as the Decoherence of Entanglement

In this view before the bifurcation, the system's components are coherently entangled and phase-locked. At the Feigenbaum point: entanglement maximises and the system reaches perfect phase correlation ($\Delta\varphi \approx \pi$). Beyond it the correlations decohere and produce chaotic emergence, an explosion of uncorrelated microstates. So chaos is the product of entanglement, its saturation and breakdown and the point where the informational coupling exceeds the coherence bandwidth. In DE, that corresponds to:

$$| \nabla S | \rightarrow 0, \nabla^2 S < 0,$$

an inflection where the curvature of information space can no longer sustain coherence, the entropic equivalent of quantum decoherence. At the Schwarzschild horizon ($r/GM = 2$), the system's inside and outside degrees of freedom are entangled so that one cannot describe the external field without its complement across the horizon. That's the essence of the holographic principle:

$$S_{BH} = \frac{kA}{4l_P^2}.$$

The DE picture interprets that surface entropy as the entanglement measure between chiral domains, the "mirror coupling" between ε and $-\varepsilon$ across the *i-axis*. So the event horizon is the macroscopic analogue of the Feigenbaum bifurcation, a boundary of maximal correlation where two informational halves, matter and mirror, are entangled reflections.

Domain	Process	Entanglement Role	DE Interpretation
Quantum	Wavefunction non-factorability	Microscopic phase coupling	Complex correlation across *i-axis*
Feigenbaum chaos	Period-doubling	Phase-locking of recursive states	Entanglement of attractors
Event horizon	Inside-outside correlation	Holographic entanglement	Mirror emergence across $\varepsilon \leftrightarrow -\varepsilon$
DE chiral universe	$\varepsilon \rightarrow -\varepsilon$ transition	Global self-entanglement	Universe reflecting through its *i-axis*

So chaos is emergent entanglement at macroscopic scale, horizons are entanglement boundaries, and chiral universes are entangled mirrors through the *i-axis.*

The Feigenbaum Map and Recursive Stability

At certain points in the recursion the real and imaginary components of the entropy gradient balance. At these points the system becomes maximally sensitive to fluctuations and transitions between regimes of behaviour. This plays a role analogous to the Feigenbaum bifurcation point in discrete dynamical systems. The comparison can be summarised schematically:

Domain	Equation	Critical Point	Interpretation
General Relativity	$1 - 2GM/r = 0$	$r/GM = 2$	Horizon pivot
Chaos theory	$x_{n+1} = f_\lambda(x_n)$	$\lambda = \lambda_c$	Bifurcation threshold
Deductive Entropy	$dS/d\varepsilon = 0$	$\varepsilon = \varepsilon_c$	Emergence pivot

Although the underlying equations differ, these points play similar structural roles: they represent boundaries where the governing dynamics reorganise. In nonlinear dynamics the transition from stable behaviour to chaos often occurs through a cascade of period-doubling bifurcations. A canonical example is the logistic map

$$x_{n+1} = f_\lambda(x_n) = \lambda x_n(1 - x_n)$$

As the control parameter λ increases, the system undergoes successive bifurcations

$$1 \rightarrow 2 \rightarrow 4 \rightarrow 8 \rightarrow \cdots$$

before entering chaotic behaviour. The spacing between these bifurcations converges geometrically according to the Feigenbaum constant

$$\delta \approx 4.6692.$$

Remarkably, this scaling is largely independent of the detailed form of the underlying map. Systems with very different microscopic dynamics nevertheless exhibit the same bifurcation scaling provided they possess a single quadratic maximum. The reason is that near the onset of chaos the dynamics become governed by a universal renormalisation operator. The Feigenbaum fixed-point map can be written schematically as

$$g(x) = -\alpha\, g\left(g\left(\frac{x}{\alpha}\right)\right),$$

where $g(x)$ is the universal map and α is the spatial rescaling constant. The essential feature is that the map reproduces its own form under recursive composition followed by rescaling. Stability therefore depends not on the detailed form of the dynamics but on whether the system preserves its structure under this renormalisation transformation. A closely related structure appears in the DE framework. Near an emergence pivot the entropy field satisfies

$$\frac{dS}{d\varepsilon} = 0,$$

with local instability indicated by negative curvature

$$\frac{d^2S}{d\varepsilon^2} < 0.$$

Expanding the entropy function locally around the pivot ε_c gives

$$S(\varepsilon) \approx S(\varepsilon_c) + \frac{1}{2}S''(\varepsilon_c)(\varepsilon - \varepsilon_c)^2 + \cdots .$$

The leading term is quadratic, which is precisely the structural condition that produces universality in Feigenbaum renormalisation. The emergence pivot therefore belongs to the same class of critical structures as the bifurcation point of a logistic map. To make the correspondence explicit, we define a recursive update operator $R(\varepsilon)$ describing the evolution of informational curvature near the pivot. Stability across scales requires that this operator preserve its form under recursive rescaling of the entropic coordinate. The DE renormalisation condition can therefore be written:

$$\mathcal{R}(\varepsilon) = \beta\,\mathcal{R}\left(\mathcal{R}\left(\frac{\varepsilon}{\beta}\right)\right)$$

where β represents the rescaling factor in ε-space. This equation is the informational analogue of the Feigenbaum fixed-point relation. In chaos theory the renormalisation operator rescales coordinate distance in state space. In DE the recursion rescales entropic distance from the emergence pivot. Both describe the preservation or loss of recursive stability under repeated composition. The correspondence can be summarised as follows.

Chaos Theory	**Deductive Entropy**
State variable x	Entropic coordinate ε
Logistic map $f(x)$	Recursive update $R(\varepsilon)$
Quadratic maximum	Emergence pivot $dS/d\varepsilon = 0$
Feigenbaum scaling	ε–renormalisation
Bifurcation cascade	Recursive emergence transitions

This comparison shows that the appearance of universality in chaos theory and the stability structure of DE recursion arise from the same mathematical principle, namely a quadratic critical point preserved under renormalisation. In chaos theory this structure produces the Feigenbaum cascade, and in the DE framework it governs the stability of informational emergence across scales. The logistic map rescales spatial coordinates, whereas the ε-recursion rescales informational distance. Despite these different interpretations, both systems follow the same renormalisation logic. The emergence pivot in DE can therefore be understood as the informational counterpart of the Feigenbaum fixed point, a boundary at which recursive dynamics reorganise and new regimes of behaviour become possible.

The Chiral Gateway

Extending this recursive stability framework to cosmology suggests that horizons act as large-scale informational boundaries in the ε-field. The following interpretation explores the consequences of that assumption. On DE's reading, there are ~4×10^{19} chiral boundaries (one per event horizon) in the observable universe dominated by stellar-mass black holes. A 2022 black-hole mass-function study (SEVN + cosmic star-formation/metallicity history) estimates $\approx 4 \times 10^{19}$ stellar-mass black holes, about ~1% of all baryons. In the Milky Way alone, population studies suggest ~10^8–10^9 stellar-mass BHs, consistent with the above scaling. Supermassive BHs (SMBHs) add at least one per massive galaxy, i.e., $\geq 10^{11}$–10^{12}, but this is a tiny correction compared with 4×10^{19}. DE twist (counting "chiral boundaries" more finely), each non-rotating BH contributes one boundary (the horizon). Most BHs rotate (Kerr), giving two horizons (outer event horizon r_+ and inner Cauchy horizon r_-), plus the ergosphere (static limit). If we treat both horizons as "chiral interfaces," the count is up to ~8×10^{19} horizon surfaces, and $\sim 1 - 2 \times 10^{20}$ if you also include ergosurface boundaries as informational interfaces. (SMBHs adjust this by <1%.) So, in DE terms the order-of-magnitude $\sim 10^{20}$ chiral interfaces in today's observable universe, with $\sim 4 \times 10^{19}$ distinct black holes setting the floor on the count.

Consider each black hole is a chiral boundary node. If the observable universe contains roughly 4×10^{19} black holes, then in DE terms we have that many active entropic mirrors, points where the emergent ε-field inverts ($\varepsilon \rightarrow -\varepsilon$). Each horizon is not a void, but an informational junction between our positive $+\varepsilon$ universe and its negative $-\varepsilon$ chiral complement. So, collectively, the universe is threaded by a vast network of phase-coupled mirrors, a lattice of informational curvature. These surfaces entangle curvature and vacuum phase, storing and releasing zero-point energy in the same way atomic orbitals quantize energy levels. In this frame dark energy may be seen as aggregate entropic pressure the accelerating expansion of the universe, not caused by an unknown "repulsive energy," but by the aggregate entropic gradient of these chiral nodes. Each horizon contributes a minute negative-pressure term to spacetime curvature. The superposition of $10^{19} - 10^{20}$ such nodes produces an average large-scale positive cosmological pressure, matching the observed Λ term:

$$\Lambda_{\text{eff}} \approx \frac{\sum_i \dot{S}_i}{V_{\text{U}}}.$$

Where $\dot{S}_i$ is the local entropy flux (positive outward) at each horizon. Because the $+\varepsilon \rightarrow -\varepsilon$ transition produces a negative curvature contribution, their ensemble effect mimics dark energy as an apparent cosmological constant arising from the sum of all chiral inversions.

The Chiral Interpretation of the Dark Sector

In the DE framework, large-scale cosmological phenomena can be reinterpreted as consequences of chiral horizon dynamics rather than as independent components. Dark energy arises as the cumulative effect of entropic backpressure generated at chiral boundaries. Each horizon acts as a local $\varepsilon \rightarrow -\varepsilon$ inversion, contributing a small effective negative-pressure term. When summed across the $\sim 10^{19} - 10^{20}$ horizon interfaces in the observable universe, these contributions produce a large-scale outward acceleration consistent with the observed cosmological constant. In this sense, dark energy can be understood as the aggregate effect of horizon-scale entropy flux rather than a separate fundamental field.

A similar reinterpretation applies to dark matter. In the DE picture, correlations between chiral boundaries generate extended curvature structures across the ε-field. These appear as filamentary connections between regions of stabilised structure. Such filaments carry curvature and influence gravitational dynamics but do not couple to electromagnetic radiation. As a result, they produce lensing and rotation-curve effects similar to those attributed to dark matter. Dark matter can therefore be interpreted as the gravitational manifestation of non-radiant curvature stored in the entangled structure of the ε-field.

The cosmic microwave background also admits a reinterpretation within this framework. Rather than being solely relic radiation from recombination, it may encode the residual microstate distribution inherited from a prior phase of the recursion. Its near-uniformity reflects a state in which entropic flux had approached equilibrium before re-emergence, while its small anisotropies correspond to the earliest deviations from that equilibrium. In this view, the CMB functions as a large-scale memory imprint of the previous chiral phase. The cosmological constant problem likewise takes a different form. In standard quantum field theory, vacuum energy contributions vastly exceed the observed value of Λ. In the DE framework, vacuum energy contributions from $+\varepsilon$ and $-\varepsilon$ domains naturally oppose one another:

$$E_{\mathrm{vac}}(\varepsilon) + E_{\mathrm{vac}}(-\varepsilon) \approx 0$$

The observed cosmological constant then reflects a small residual asymmetry between the two chiral domains rather than a large absolute energy density. Taken together, these effects suggest that the so-called "dark sector" may not consist of independent unknown components, but instead reflects the large-scale consequences of the universe's chiral recursion structure.

Structural Summary

Phenomenon	Standard View	DE Interpretation	Mechanism
Black holes	Information-trapping singularities	Chiral inversion nodes	$\varepsilon \rightarrow -\varepsilon$ transitions
Dark energy	Cosmological constant	Aggregate horizon backpressure	Entropy flux across boundaries

Phenomenon	Standard View	DE Interpretation	Mechanism
Dark matter	Non-luminous mass	Non-radiant curvature structures	ε-field correlations
CMB	Relic radiation	Residual microstate distribution	Prior-phase equilibrium imprint
Λ / vacuum energy	Zero-point energy mismatch	Residual chiral asymmetry	Incomplete cancellation

In this framework, the Big Bang becomes a phase transition across the chiral interface. The collective evolution of prior horizons drives the system toward a state of near symmetry, after which a new phase of emergence begins.

$$(\varepsilon, -\varepsilon) \rightarrow (+\varepsilon')$$

This represents a reset of phase orientation rather than creation ex nihilo, consistent with a continuous recursive cosmology.

Horizons and the Fractal Symmetry of Emergence

Within the DE framework each black hole horizon can be viewed as a local mirror boundary in the ε field. Across the horizon the informational phase reverses sign, creating a local $\varepsilon \rightarrow -\varepsilon$ transition. In this sense a black hole resembles a microscopic version of the global mirror transition associated with the cosmological Big Bang. This suggests a fractal symmetry between the small and large scales of the recursion:

$$\text{micro (black hole)} \iff \text{macro (cosmos)}.$$

Both obey the same entropy oscillation law but operate at vastly different phase scales. In conventional cosmology inflation is produced by an assumed scalar potential $V(\phi)$. In the DE picture the potential arises naturally from the curvature energy of the entropic field itself,

$$V(\varepsilon) = \frac{1}{2}k(1 - \cos\, k\varepsilon).$$

The gradient of this potential drives the transient super-expansion when ε crosses the mirror plane. Inflation is therefore the wave of informational realignment following the mirror transition rather than the action of an independent field.

Cyclic Entropy and the Mirror Node

Standard cosmology interprets the second law of thermodynamics within a one-directional time frame. Entropy increases until gradients vanish, leading to the familiar "heat death" scenario. This conclusion assumes a non-chiral universe with a single direction of informational flow. The DE framework relaxes this assumption, as entropy becomes an oscillatory quantity associated with alternating chiral domains,

$$S(\varepsilon) = S_0 + A\cos\,(k\varepsilon).$$

When the entropy gradient approaches zero the system does not terminate. Instead it approaches the mirror node where

$$\varepsilon \to -\varepsilon,$$

and the direction of informational flow reverses. Over a complete cycle the net informational change therefore vanishes,

$$\int_{+\varepsilon}^{-\varepsilon} dS = 0.$$

From this perspective the universe behaves as a conservative informational system rather than a dissipative one. What appears locally as entropy growth is one phase of a larger oscillatory cycle in which emergence alternates between complementary chiral domains.

Time, Recursion, and the $\varepsilon - \Upsilon$ Conjugate

In the DE framework time becomes a projection of informational emergence. The primary ordering variable is the entropic distance ε, while structural commitment is measured by its conjugate quantity Υ. These two quantities form a natural pair,

$$\Xi = (\varepsilon, \Upsilon)$$

where ε measures informational displacement from equilibrium and Υ measures the degree of committed structure.

Time arises as the observable ordering of this pair. In continuous regimes the relationship is approximately

$$t \propto \varepsilon,$$

so the perceived arrow of time reflects the direction of the entropic gradient. When the gradient reverses, the experienced direction of time reverses as well. Globally, the universe therefore has no absolute beginning or end. Instead it consists of alternating regions in which emergence proceeds along opposite orientations of the ε-field.

Entropy as Predictive Compression

In classical thermodynamics entropy is usually interpreted as a measure of disorder or lost information. In the DE framework entropy instead measures predictive compression, the degree to which the system can infer its own state across its configuration space. This can be written schematically as

$$S = k \ln P_{\text{predictive}}$$

where $P_{\text{predictive}}$ represents the probability that the system's inferred structure correctly predicts its future states. Entropy therefore measures how efficiently the universe encodes its own dynamics. Across a complete chiral cycle the informational balance satisfies

$$\int_{+\varepsilon}^{-\varepsilon} dS = 0$$

so the universe remains globally conservative in information even though local observers perceive monotonic entropy increase.

Cyclic Emergence and Informational Refinement

The ε-field exists in two complementary chiral domains, so that their entropy functions form complex conjugates,

$$S_- = S_+^*.$$

A small phase difference between the two domains accumulates over successive cycles. This produces a statistical bias toward greater informational compression. If

N denotes the number of cycles and Π denotes a parsimony index measuring informational efficiency, then

$$\frac{d\Pi}{dN} > 0.$$

Each cycle therefore inherits informational constraints from the preceding one. Black hole horizons act as transmission channels for these correlations, allowing the microstate structure of one cosmic phase to seed the next with non-random priors. Structure formation does not begin from a completely unconstrained state, instead each cycle begins with partially optimized informational structure inherited from its predecessor.

Minimal Chiral Cosmology

Many cosmological models propose a multiverse consisting of many disconnected universes. The DE framework predicts a simpler structure, where the informational field exists in two complementary chiral domains,

$$\Psi(-\varepsilon) = \Psi(+\varepsilon)^*.$$

These domains are not separate universes but phase reflections of the same informational manifold. Apparent branching corresponds to local phase separation within the same ε-field rather than the creation of independent realities. The universe therefore adopts the minimal symmetry required to encode its information, a dual chiral structure $(\varepsilon, -\varepsilon)$.

The DE Cosmological Recursion Law

All of the preceding ideas can be summarised by a single recursive relation governing the evolution of informational structure:

$$\Xi_{n+1} = \mathcal{R}(\Xi_n)$$

with

$$\Xi = (\varepsilon, \Upsilon).$$

Near a critical pivot the recursion takes the renormalised form

$$\Xi_{n+1} = \alpha^{-1}\ \mathcal{R}(\Xi_n)$$

which mirrors the Feigenbaum renormalisation operator. Here α represents the universal scaling factor controlling the contraction of accessible informational states. This equation captures the essential architecture of DE cosmology:

- ε tracks informational displacement (the observable arrow of time)
- Υ measures committed structure
- the recursion operator $\mathcal{R}$ generates emergence across scales
- renormalisation preserves stability through universal scaling.

Continuous physical laws emerge when the recursion is sampled smoothly in ε, producing the spacetime curvature of general relativity. Discrete physical laws emerge when the recursion is sampled stepwise, producing the quantised structures of particle physics. Cosmology occupies the interface between these regimes. The entire DE framework can therefore be summarized by the evolution of the $\varepsilon - \Upsilon$ pair under recursive renormalisation:

$$(\varepsilon, \Upsilon)_{n+1} = \alpha^{-1}\, \mathcal{R}\, (\varepsilon, \Upsilon\,)_n$$

From this single rule arise:

- oscillatory dynamics in continuous regimes
- bifurcation cascades in discrete regimes
- horizon pivots in gravitational systems
- and the large-scale evolution of the cosmos.

The universe evolves because informational structure recursively compresses toward greater parsimony while preserving its underlying symmetry across the chiral domains of the ε-field.

Informational Geometry and the Emergence of Spacetime

The Deductive Entropy framework begins with a simple premise that emergence is governed by informational dynamics rather than by spacetime itself. The fundamental variables are the entropic displacement ε and the degree of committed informational structure Υ. These quantities form a conjugate pair describing the evolution of the informational state of the system,

$$\Xi = (\varepsilon, \Upsilon).$$

Entropic displacement measures how far the system lies from equilibrium, while structural commitment measures the degree to which informational constraints have stabilized into persistent structure. The two quantities cannot vary independently. Increasing informational displacement alters the amount of committed structure the system can sustain, and increasing commitment constrains the directions available for further informational exploration. Their relationship mirrors the canonical conjugate pairs familiar from physics, such as position and momentum in classical mechanics or field amplitude and momentum density in field theory. In the DE framework emergence therefore proceeds through the coupled evolution of ε and Υ.

The Informational Action

Because ε and Υ behave as conjugate variables, their dynamics can be expressed through an action principle. A natural informational action takes the form

$$\mathcal{A} = \int (\Upsilon\, d\varepsilon - H(\varepsilon, \Upsilon)),$$

where $H(\varepsilon, \Upsilon)$ represents the informational Hamiltonian governing the evolution of the system. Stationary variation of this action produces evolution equations structurally identical to Hamilton's equations,

$$\frac{d\varepsilon}{d\tau} = \frac{\partial H}{\partial \Upsilon},$$

$$\frac{d\Upsilon}{d\tau} = -\frac{\partial H}{\partial \varepsilon}.$$

Here τ represents the ordering parameter of the recursion. In observational regimes this parameter appears as time. The familiar temporal ordering of physical processes therefore emerges as a projection of informational evolution along the ε-axis. Time is not fundamental but the observable ordering of informational displacement.

The Informational Metric

Once the informational state space is identified, it becomes natural to define a metric on this manifold. The coordinates of the informational manifold may be written

$$\Xi^A = (\varepsilon, \Upsilon, \theta, \phi),$$

where (θ, ϕ)describe the orientation of informational phase. A simple informational line element is

$$d\sigma^2 = d\varepsilon^2 + \lambda^2 d\Upsilon^2 + R^2(\varepsilon, \Upsilon)(d\theta^2 + \sin^2\theta \, d\phi^2).$$

This metric does not yet describe spacetime, rather it describes the geometry of informational emergence itself. Curvature arises because the effective radius $R(\varepsilon, \Upsilon)$ depends on the informational state of the system. Entropic displacement and structural commitment therefore deform the geometry of the orientation sphere. In this picture gravity is not introduced as an external interaction, it arises because informational dynamics curve the manifold of possible states.

Informational Curvature and Emergent Gravitation

Once a metric is defined, the standard machinery of differential geometry follows automatically. The informational manifold possesses a Riemann curvature tensor, a Ricci tensor, and a scalar curvature,

$$\mathcal{R}^A{}_{BCD}, \mathcal{R}_{AB}, \mathcal{R}.$$

From these quantities we construct the informational analogue of the Einstein tensor,

$$\mathcal{G}_{AB} = \mathcal{R}_{AB} - \frac{1}{2} g_{AB} \mathcal{R}.$$

Gradients in the informational variables generate an informational stress tensor,

$$\mathcal{T}_{AB}^{(\Xi)} = \nabla_A \varepsilon \nabla_B \varepsilon + \mu \nabla_A \Upsilon \nabla_B \Upsilon - g_{AB} \mathcal{L}_{\Xi}.$$

The natural field equation for the informational manifold is therefore

$$\mathcal{G}_{AB} = \kappa \mathcal{T}_{AB}^{(\Xi)}.$$

This equation governs the curvature of informational space. In the regime where the informational dynamics are sampled smoothly, the projection of these tensors into observable spacetime reproduces the structure of the Einstein field equations. General relativity therefore appears as the continuous limit of informational curvature. Spacetime is not fundamental, it is the low-resolution projection of the deeper informational manifold.

Horizons and the Failure of Spacetime Description

Within this framework gravitational collapse acquires a new interpretation. A spacetime singularity represents the point where the spacetime description ceases to be valid. The underlying informational manifold, however, remains regular. Black hole horizons correspond to surfaces where the spacetime projection becomes singular while the informational geometry remains well defined. Collapse therefore signals the breakdown of spacetime coordinates rather than the termination of physical structure. The informational state simply passes through a phase inversion

$$\varepsilon \rightarrow -\varepsilon.$$

This inversion connects two complementary orientations of the informational manifold.

The Chiral Structure of Informational Space

The informational manifold possesses a natural chiral structure. The two domains

$$+\varepsilon \text{ and } -\varepsilon$$

represent opposite orientations of informational emergence. These domains are not independent universes but complementary halves of the same informational structure. Their entropy functions form complex conjugates,

$$S_- = S_+^*.$$

The transition between them occurs at the surface where

$$\varepsilon = 0.$$

This surface functions as a mirror boundary between the two orientations of informational time. The universe therefore possesses a minimal dual symmetry rather than a collection of independent universes.

Penrose Geometry and the Chiral Information Sphere

Remarkably, a geometric structure encoding exactly these features already exists within relativistic geometry. Penrose's Riemann sphere represents the projective space of spinor directions,

$$S^2 = \mathbb{CP}^1.$$

Every point on this sphere corresponds to a null direction in spacetime. The poles represent opposite complex

orientations of the spin frame, while the equator corresponds to real null directions.
The sphere therefore separates naturally into two hemispheres representing opposite frequency orientations of the spinor field. Positive-frequency modes occupy one hemisphere, while negative-frequency modes occupy the other. Crossing the equator corresponds to reversing the orientation of time in the spinor description.

This structure mirrors precisely the chiral organization of the informational manifold. Within the DE interpretation the polar axis of the sphere corresponds to the ε-axis of informational emergence. The north and south poles represent the two orientations of informational time,

$$+\varepsilon \text{ and } -\varepsilon.$$

The equator represents the transition surface where informational phase reverses. Penrose's sphere therefore provides a geometric representation of the DE chiral universe. What appears in relativistic geometry as the sphere of spinor directions is revealed, from the informational perspective, to be the minimal structure capable of encoding orientation, phase, and causal direction within a single manifold.
The Riemann sphere thus becomes the simplest possible carrier of informational chirality.

The Informational Origin of Physical Law

Seen in this way the major structures of physics arise from different projections of the same informational geometry. Continuous trajectories on the manifold produce the smooth curvature of spacetime described by general relativity. Discrete sampling of the recursion produces the branching structures characteristic of nonlinear dynamics and quantum interactions. Horizon boundaries represent the points where the spacetime projection fails and informational phase inverts. The chiral information sphere provides the minimal geometry organizing all of these phenomena. Physics therefore emerges not from spacetime itself but from the deeper geometry of informational recursion operating on the (ε, Υ) manifold. Spacetime, gravity, and quantum structure all appear as different observational regimes of the same underlying informational dynamics.

Cohomology as the Engine of Fields and Particles

In twistor theory physical fields are not defined primarily as functions on spacetime. Instead they arise from cohomology classes on twistor space. For a massless field of helicity h, the corresponding object is

$$H^1(PT, \mathcal{O}(2h-2)).$$

Elements of this cohomology group are holomorphic structures on projective twistor space. When the Penrose transform is applied, these classes generate the familiar spacetime fields such as Maxwell, Weyl, or Dirac fields. From this perspective a physical field is not a local quantity defined point-by-point in spacetime. It is an equivalence class of holomorphic configurations on twistor space. Particles correspond to stable, nontrivial cohomology classes. Interactions arise through algebraic operations such as cup products or through relationships encoded in exact sequences of sheaves. The power of cohomology lies in the fact that it captures global information that cannot be eliminated by local deformations. Cohomology therefore acts as a filter that removes transient fluctuations while preserving structures that remain stable under continuous transformations. In this sense cohomology functions as a form of topological memory within a complex analytic manifold.

Cohomology and the Entropic Landscape

The Deductive Entropy framework interprets these cohomological structures in informational terms. The $\varepsilon - \Upsilon$ manifold describes the landscape of informational emergence, where

- ε measures entropic displacement
- Υ measures accumulated structural complexity.

Within this landscape stable attractors correspond naturally to nontrivial cohomology classes.

Twistor Cohomology	**DE interpretation**
Nontrivial cohomology class	Stable attractor in $\varepsilon - \Upsilon$ space
Closed but not exact form	Low-entropy configuration with memory
Holomorphic equivalence	Minimal informational representation
Cohomology dimension	Degrees of freedom of structured solutions
Exact sequence	Recursive relations on the DE manifold

From this perspective cohomology appears whenever the universe stores information in a form that cannot be reduced to a trivial boundary. Cohomological structures survive precisely because they resist dissipation. They represent configurations that remain invariant under recursive uncertainty filtering. In the DE framework these become the informational attractors that manifest as persistent physical structures.

Chirality and Entropic Orientation

Twistor theory distinguishes helicity states through separate cohomology classes:

$$\mathcal{O}(2h-2) \text{ vs } \mathcal{O}(-2h-2).$$

This algebraic split reflects the existence of two complementary orientations in the complex structure. Within DE this separation corresponds directly to the chirality axis of informational flow. The chirality parameter ρ marks the orientation of informational emergence:

- positive helicity $\leftrightarrow \rho > 0$
- negative helicity $\leftrightarrow \rho < 0$

Cohomology therefore records global informational separation between distinct attractor classes.
Two configurations belonging to the same cohomology class differ only by exact forms and thus correspond to small entropic displacement ε. Configurations belonging to different classes occupy separate basins of the entropic landscape and therefore correspond to large entropic separation. Matter types thus appear as stable attractors within distinct informational basins.

Phase Memory and the Imaginary Axis

In Penrose's twistor construction holomorphic structures naturally carry complex phase information. These phases are preserved under cohomological mappings. In the DE framework the imaginary axis represents the memory axis of the informational manifold. Cohomology preserves global phase relationships across this axis, allowing the system to retain structural information across recursive updates. This phase preservation explains why cohomology classes behave as stable entities. They encode information in a form that survives repeated transformations of the underlying manifold. Cohomology therefore acts as the algebraic representation of DE's phase-memory mechanism.

The Informational Origin of the Speed of Light

In conventional physics the speed of light appears as a fundamental constant built into spacetime. In the DE framework this constant emerges from the relationship between informational displacement and geometric separation. The entropic coordinate ε tracks the resolution of uncertainty along the informational axis. Spacetime geometry emerges as a representation of how that informational structure organizes itself. Along special trajectories known as null paths, information propagates with maximal efficiency. These paths correspond to the propagation of light. There exists a fixed conversion factor between entropic displacement and geometric distance along such paths. If s denotes the spacetime interval, the relationship can be written

$$\frac{ds}{d\varepsilon}\Big|_{\text{null}} = c.$$

Here c appears not as a property of light itself but as the conversion factor between informational distance and geometric distance along optimal information trajectories. The inverse form

$$\frac{d\varepsilon}{ds}\Big|_{\text{null}} = \frac{1}{c}$$

expresses the entropic cost of supporting a given geometric displacement. The speed of light therefore emerges as the universal exchange rate linking information and geometry.

Mass, Energy, and Entropic Geometry

Einstein's relation

$$E = mc^2$$

can now be rewritten using the informational interpretation of c:

$$E = m\left(\frac{ds}{d\varepsilon}\right)^2.$$

This expression reveals that mass, energy, and entropy represent different manifestations of the same informational geometry. Rearranging gives

$$m = E\left(\frac{d\varepsilon}{ds}\right)^2.$$

When informational displacement occurs slowly relative to geometric separation, mass appears large. Mass therefore corresponds to information that resists propagation across the entropic manifold. Energy, by contrast, represents informational structure that propagates freely. Thus energy corresponds to entropy in motion and mass corresponds to entropy stabilized into persistent structure. The constant c^2 expresses the square of the informational conversion factor linking these two regimes.

Informational Geometry as the Foundation of Physics

Seen from this perspective the constants of physics are not arbitrary parameters imposed on spacetime. They arise from deeper relationships governing informational emergence. Cohomology provides the classification of stable informational structures. Twistor space supplies the complex analytic framework in which these structures live. The $\varepsilon - \Upsilon$ manifold describes the entropic landscape through which informational evolution proceeds. Within this unified picture:

- fields correspond to cohomology classes
- particles correspond to stable informational attractors
- the speed of light expresses the conversion between informational and geometric displacement
- mass and energy represent complementary modes of informational structure.

The laws of physics therefore appear as the geometric expression of deeper informational principles governing the recursive organization of the universe.

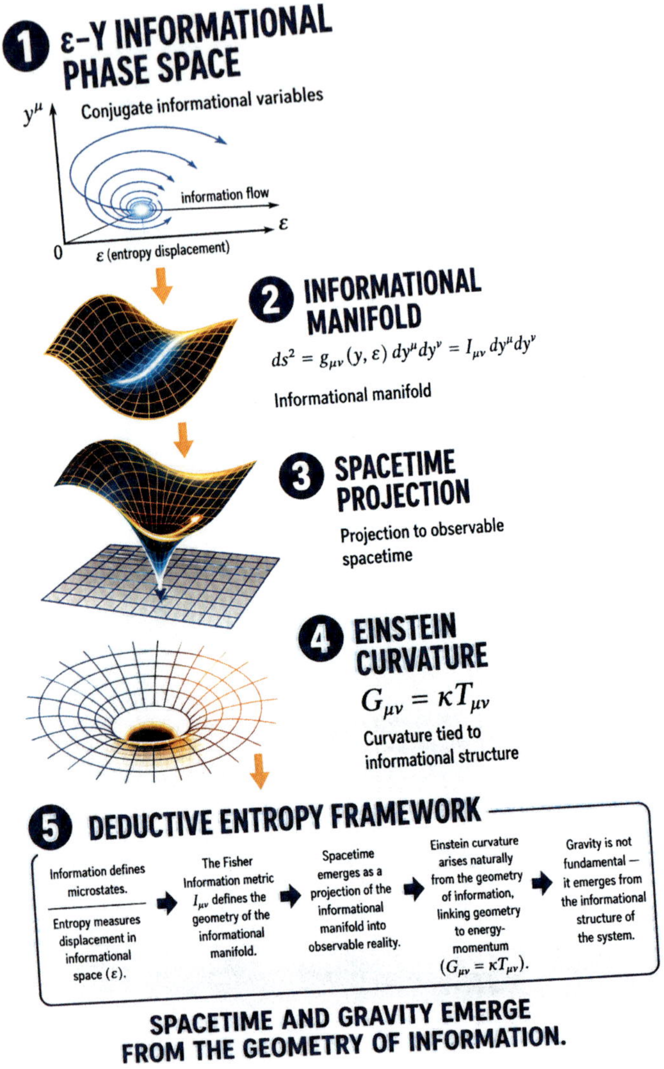

Emergence of spacetime from informational geometry.

The Deductive Entropy framework begins with the conjugate informational variables ε (entropic displacement) and Υ (structural commitment). Their recursive dynamics generate an informational manifold with metric $d\sigma^2$. In the continuous limit this geometry projects into the spacetime metric of general relativity. Einstein curvature therefore appears as the observable projection of deeper informational dynamics.

Appendix; Dynamics in Recursion Space

Phase progression generates structural memory through stabilisation. The rate of accumulation is governed by

$$\frac{d\Sigma}{d\varepsilon} = \Upsilon(\varepsilon)$$

ε determines ordering direction, Υ determines which phase relationships persist, and Σ records the accumulated retained phase. Introduce an informational potential

$$V(\Sigma) = \frac{1}{2}k\Sigma^2$$

and define a recursion Lagrangian

$$L = \frac{1}{2}\mu\left(\frac{d\Sigma}{d\varepsilon}\right)^2 - \frac{1}{2}k\Sigma^2$$

The Euler–Lagrange equation gives

$$\frac{d^2\Sigma}{d\varepsilon^2} = -\Omega^2\Sigma$$

with $\Omega^2 = k/\mu$. This describes a harmonic oscillator in recursion space. The conserved quantity

$$E_\Sigma = \frac{1}{2}\mu\left(\frac{d\Sigma}{d\varepsilon}\right)^2 + \frac{1}{2}k\Sigma^2$$

represents the balance between phase accumulation (conversion of ordering into retained structure) and stored structural curvature (displacement from the stabilisation pivot). The pivot occurs at $\Sigma = 0$, where retained phase vanishes and the direction of accumulation reverses, and what appears as oscillation is the alternating retention and release of phase around a stabilisation point.

1. Core Definitions

Let ε index progression through ordering space. Phase is the primary variable of emergence, so structural memory arises only from the accumulation of phase that survives stabilisation. Define the retained structural memory

$$\Sigma(\varepsilon) = \int_{\varepsilon_i}^{\varepsilon_f} \Upsilon(\varepsilon')\, d\varepsilon'[\Sigma] = \mathrm{J\backslash cdotps}$$

Σ is the accumulated record of stabilised phase. Define the observable phase

$$\theta(\varepsilon) = \frac{\Sigma(\varepsilon)}{\hbar}$$

Planck's constant sets the scale at which accumulated phase becomes observable. Postulate the resolution floor

$$\Delta\Sigma_{\mathrm{min}} = \hbar$$

so that

$$\Delta\Sigma = n\hbar, \oint d\Sigma = 2\pi n\hbar$$

Quantum structure therefore reflects the discreteness of retained phase accumulation.

2. Phase Flow

Phase accumulation rate is defined along ε:

$$\frac{d\Sigma}{d\varepsilon} = \Upsilon(\varepsilon)$$

This is the key identity:

$\varepsilon \rightarrow$ orders progression
$\Upsilon \rightarrow$ determines what survives
$\Sigma \rightarrow$ accumulates the result

$$J_\Sigma(\varepsilon) \equiv \frac{d\Sigma}{d\varepsilon}$$

Then:

$J_\Sigma > 0$: forward accumulation
$J_\Sigma < 0$: reverse accumulation
$J_\Sigma = 0$: pivot

3. Harmonic Oscillator in Recursion Space

Introduce an informational potential

$$V(\Sigma) = \frac{1}{2}k\Sigma^2$$

Define the recursion Lagrangian

$$L_{\mathrm{DE}} = \frac{1}{2}\mu\left(\frac{d\Sigma}{d\varepsilon}\right)^2 - \frac{1}{2}k\Sigma^2$$

The Euler–Lagrange equation gives

$$\frac{d^2\Sigma}{d\varepsilon^2} = -\Omega^2\Sigma \text{with} \Omega^2 = \frac{k}{\mu}$$

This is the DE harmonic oscillator in recursion space. The conserved quantity

$$E_\Sigma = \frac{1}{2}\mu\left(\frac{d\Sigma}{d\varepsilon}\right)^2 + \frac{1}{2}k\Sigma^2$$

represents:

- phase accumulation rate (flow through recursion)
- stored structural memory (distance from pivot)

The pivot occurs at

$$\frac{d\Sigma}{d\varepsilon} = 0$$

where accumulation halts and reverses

4. Mapping to the Schwarzschild Envelope

Define

$$x = \frac{r}{GM}, x \geq 2$$

$$\varepsilon_{\text{env}}(x) = \pm\sqrt{1 - \frac{2}{x}}$$

An envelope on phase flow.

$$\frac{d\Sigma}{d\varepsilon} = J_0\, \varepsilon_{\text{env}}(x)$$

At the pivot $x = 2$:

$$\varepsilon_{\text{env}} = 0 \Rightarrow \frac{d\Sigma}{d\varepsilon} = 0$$

This matches the oscillator turning point. So:

- GR square-root → envelope on phase accumulation
- $\pm$ branch → chirality
- horizon → pivot

5. Role of $\hbar$ and π

Because

$$\theta = \frac{\Sigma}{\hbar}$$

the oscillator becomes

$$\frac{d^2\theta}{d\varepsilon^2} = -\Omega^2\theta$$

Planck's constant defines the minimum increment of retained phase. A chirality bifurcation satisfies

$$\Delta\theta = \pi \Rightarrow \Delta\Sigma = \pi\hbar$$

Thus:

- $\hbar$ → resolution scale
- π → bifurcation threshold

6. ϒ as Stabilisation Operator

Decompose phase sectors

$$\mid \psi\rangle = \sum_m c_m \mid \psi_m\rangle, \theta_m = \frac{\Sigma_m}{\hbar}$$

Define ϒ as projection onto closed phase classes

$$\Upsilon = \sum_{m \in C} \mid \psi_m \rangle \langle \psi_m \mid$$

Closure condition:

$$\theta_m(\varepsilon + \Delta\varepsilon) - \theta_m(\varepsilon) = 2\pi n_m$$

Then

$$\mid \psi \rangle \mapsto \frac{\Upsilon \mid \psi \rangle}{\parallel \Upsilon \mid \psi \rangle \parallel}$$

- Υ selects persistence.
- SU(2): bifurcation without closure
- SU(3): closure → stable entanglement

In the two-branch regime associated with SU2 symmetry, correlations form but cannot persist. The system continually builds relational tension but lacks a stable attractor in which correlations can close. At the SU3 transition the geometry changes fundamentally, as three internal branches become available and the system can form a closed triad. Gluon self-interaction introduces feedback, and confinement prevents correlations from dispersing. At this moment in the universe correlations cannot leak away, producing the first persistent entanglement domain where complexity becomes non-zero, and from this point onward every higher level of structure inherits the same stability rule.

Chapter 10: Scale

One of the deepest puzzles in cosmology is why the universe contains so much stable structure at all. From elementary particles to galaxies, reality appears finely balanced between chaos and rigidity. Matter persists, atoms form, stars ignite, and life emerges, yet the laws governing these processes are remarkably simple. The question is therefore not merely how structure forms, but why the universe permits such a vast hierarchy of stable organisation. Within the DE framework, the answer to this question can be derived from a single global constraint imposed at the largest scale of the universe, which is the cosmological constant Λ. The cosmological constant sets the curvature scale of spacetime and determines the largest region over which correlations can remain coherent. In an accelerating universe this scale corresponds to the de Sitter horizon

$$L_\Lambda \sim \sqrt{\frac{1}{\Lambda}}$$

which represents the maximum distance over which causal structure can persist. At the opposite extreme lies the Planck length l_P, the smallest scale at which spacetime retains operational meaning, and between these two limits lies the entire dynamic range available for the formation of structure. To express this range in a dimensionless way, the cosmological constant can be written in Planck units

$$\lambda \equiv \Lambda l_P^2$$

Observations show that this number is extraordinarily small,

$$\lambda \sim 10^{-122}.$$

This tiny value implies that the universe possesses an enormous dynamic range between its smallest and largest meaningful scales. The total hierarchy depth available for structure formation is therefore

$$R \equiv \frac{L_\Lambda}{l_P} \sim \frac{1}{\sqrt{\lambda}}.$$

For the observed value of Λ this ratio is approximately $R \sim 10^{61}$. This number measures the total geometric and informational range through which uncertainty can be converted into persistent structure. Across mathematics, physics, biology, and complex systems, chaotic systems stabilise through feedback at discrete thresholds. When order emerges through repeated self-interaction, stability is discontinuous and appears at specific transition points where correlations begin to reinforce themselves. These transitions accumulate geometrically and their spacing is governed by a universal scaling constant. In nonlinear systems this constant is the Feigenbaum constant

$$\alpha \approx 2.503.$$

Feigenbaum's constant describes the geometric contraction that occurs as systems approach stability through recursive feedback. In the DE framework, each stabilisation threshold corresponds to a qualitative transition where many degrees of freedom collapse into a single coherent structure. Below such a threshold correlations flicker and dissolve, at the threshold feedback closes on itself, and above it correlations persist and become memory. Particles, atoms, chemical bonds, living cells, organisms, and gravitationally bound systems can all be understood as successive stabilisation events in such a recursive hierarchy. If each stabilisation contracts the effective degrees of freedom by the same universal factor α, then the full hierarchy of the universe must satisfy

$$R \approx \alpha^N$$

where N is the total number of stabilisation thresholds available between the smallest and largest physical scales. Taking logarithms gives

$$N \approx \frac{\ln R}{\ln \alpha}.$$

Substituting the cosmological depth

$$R \sim \frac{1}{\sqrt{\lambda}}$$

yields

$$N \approx \frac{\ln\,(1/\sqrt{\lambda})}{\ln \alpha}.$$

Using the observed values

$$\lambda \sim 10^{-122}, \alpha \approx 2.503$$

we obtain

$$\ln\,(1/\sqrt{\lambda}) \approx \frac{1}{2} \cdot 122 \ln\ 10 \approx 140.35$$

and therefore

$$N \approx \frac{140.35}{0.916} \approx 153.$$

The deeper point is simply that the universe possesses a large but finite hierarchy depth, and that the value derived from cosmology lies naturally in the range of a few hundred stabilisation thresholds. The constants that appear in fundamental physics can then be understood within DE as

conversion interfaces between different aspects of the same underlying process. The Planck constant $\hbar$ governs the relationship between phase and energy. Phase itself is dimensionless, but when phase evolution becomes physically meaningful it carries an energetic cost. The Planck constant therefore converts phase evolution into physical action. In DE language, $\hbar$ provides the link between phase rotation on the $U(1)$ fibre and the entropic coordinate ε that measures irreversible uncertainty resolution. The speed of light c performs a complementary role, where c acts as the conversion factor that links entropic resolution to energy propagation along null informational paths. Together these constants form a chain of correspondence
order 10^2 stabilisation thresholds

$$\text{phase} \rightarrow \text{entropy } (\hbar) \rightarrow \text{energy and geometry } (c).$$

They therefore represent the exchange rates that allow different physical descriptions of the same informational process to communicate with one another. Because the cosmological constant limits the largest scale of coherent correlation and the Planck length defines the smallest meaningful scale, the universe contains a finite dynamic range within which recursive stabilisation can occur. Each stabilisation threshold consumes part of this range by locking uncertainty into persistent correlations. As thresholds accumulate, the available entropic landscape gradually fills with structure. The hierarchy therefore resembles a ladder of stabilisation events;

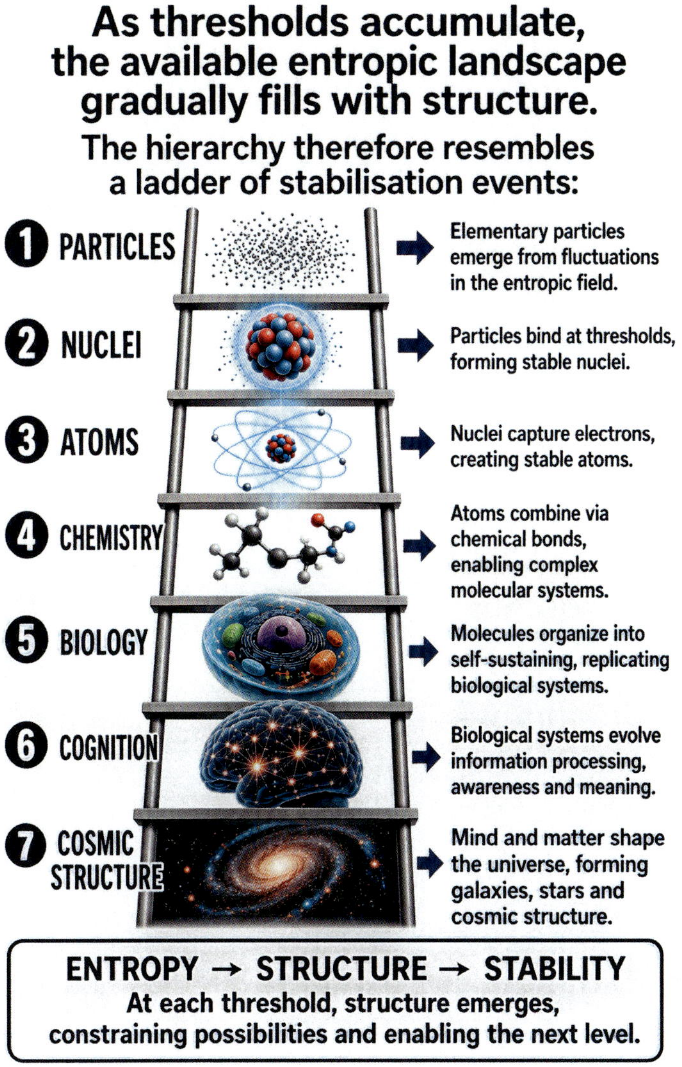

Each rung represents a qualitative change in the capacity of matter to store and transmit memory. Because the recursion depth is finite, the hierarchy of structure must also be finite. The universe therefore produces a rich but bounded spectrum of stable forms rather than an infinite cascade of ever finer organisation. In this view the universe is expanding because it is progressively converting uncertainty into memory. Each stabilisation threshold represents a 'lesson' learned by the cosmos, a transition where correlations that once flickered into existence become permanently encoded into structure. So that particles remember symmetries, atoms remember electromagnetic stability, life remembers its chemistry, and even civilisations remember information. Eventually however the recursion budget is exhausted. At that point new stable forms can no longer appear, so although expansion continues novelty fades. What remain are the deepest and most persistent memories of the universe, and those memories are stored in the most stable structures known, namely black holes, the final repositories of correlation once no further stabilisation thresholds can be crossed. In this sense the universe can be understood as a self-organising system that records its own history. Structure exists because memory persists, accumulating step by step across thresholds of recursion until the available cosmological depth is exhausted.

Although the calculation suggests that the universe may contain of the order 10^2 stabilisation thresholds, only a small fraction of these correspond to the familiar physical transitions that shape the visible world. The earliest thresholds occur at extremely small scales where quantum fields condense into stable particles. Later thresholds appear as new forms of organisation emerge from those particles. The first part of the recursion ladder therefore aligns naturally with the known hierarchy of physical structure:

Each transition represents a qualitative stabilisation event. From the DE perspective these transitions correspond to points where uncertainty collapses into durable memory, and once the threshold is crossed the system no longer explores the previous configuration space. Instead it operates within a new regime defined by the stability of the newly formed structure. The early universe therefore climbed the first several rungs of this ladder rapidly as fundamental particles, atoms, and chemistry emerged. Later rungs appear more sparsely as increasingly complex feedback loops become necessary for stabilisation.

Whilst therefore the universe possesses a recursion budget of roughly one hundred and fifty stabilisation thresholds, the known physical hierarchy occupies only a small fraction of this depth. Most of the ladder lies either in domains that are difficult to observe directly or in regimes that have not yet been explored. This should not be surprising as the first few thresholds correspond to well-studied transitions in particle physics and chemistry. Higher thresholds involve increasingly complex feedback systems whose stability depends on environmental conditions, evolutionary dynamics, and large-scale structure formation. The important point is that cosmology appears to permit a large but finite number of such transitions. The universe therefore behaves like a recursive system with a limited number of stabilisation opportunities. Each time a new threshold is crossed, correlations become locked into memory and a new level of organisation appears.

Conformal–Entropic Correspondence (CEC)

Structure formation across physics often appears as a recursive process in which stable patterns emerge through repeated contraction and reorganisation of degrees of freedom. Renormalisation theory describes this behaviour formally as a system is coarse-grained, only certain correlations survive and reappear at larger scales. In the DE framework, this process can be interpreted as the progressive compression of uncertainty into stable informational structure. This perspective suggests a simple geometric picture, where each recursive step of structure formation can be represented as a transformation in a complex information space that performs three operations simultaneously:

1. **Radial contraction**, reducing the number of independent degrees of freedom.
2. **Phase rotation**, preserving correlations and preventing collapse into trivial equilibrium.
3. **Projection onto a scale-invariant structure**, allowing patterns to repeat across scales.

Together these operations generate a logarithmic spiral in state space and repeated application of the transformation drives the system toward a stable attractor where correlations persist across scales. In nonlinear dynamics, the approach to stability is governed by universal scaling laws. The Feigenbaum constant

$$\alpha \approx 2.503$$

describes the contraction of state-space amplitudes during successive bifurcations in recursive systems. The total number of such conformal contractions is not arbitrary but bounded by the cosmological scale ratio $R \sim 1/\sqrt{\lambda}$, linking the local contraction factor α to the global recursion budget set by λ. Feigenbaum α appears in a wide range of systems approaching the onset of chaos. Within the DE interpretation, this contraction reflects the progressive reduction of uncertainty. Each recursion step compresses the range of possible configurations while preserving the correlations that proved stable in earlier iterations. Without rotation the contraction would collapse directly to a trivial fixed point, but with an appropriate phase shift the trajectory instead spirals toward a stable attractor. This spiral convergence is the geometric origin of self-similar structure.

Conformal Geometry and Twistor Space

A natural mathematical arena for representing transformations that combine scale and orientation is conformal geometry. Conformal maps preserve angles while allowing local rescaling, making them well suited to describe systems where patterns repeat across different scales. Twistor theory provides a particularly elegant formulation of conformal geometry. In twistor space, physical configurations are represented by families of null directions associated with light propagation. Twistor coordinates therefore encode both scale and orientation in a unified complex structure. If recursive structure formation is interpreted as a conformal contraction with phase transport, then twistor space provides a natural setting in which this process can be represented. Each recursion step can be written schematically as

$$Z \mapsto \frac{1}{\alpha} e^{i\phi} Z,$$

where Z represents a point in the complex informational manifold encoding both amplitude (uncertainty) and phase (correlation structure), while conformal relations between directions are preserved through projective transformations acting on the twistor fibre. Under repeated application of this transformation, the system approaches a fixed twistor configuration representing a stable correlation structure. Self-similarity emerges naturally because conformal maps preserve the relationships between patterns even as their scale changes.

Entanglement as Correlation Geometry

In quantum theory, correlations between subsystems are measured by entanglement entropy. Within the conformal-entropic correspondence, entanglement can be viewed geometrically as the preservation of phase relationships between different regions of the informational manifold. Two interacting subsystems can therefore be represented as curves in the underlying complex geometry whose relative phase determines whether their interference is constructive or destructive. Stable correlations correspond to configurations where these phase relations remain invariant under recursive contraction. This interpretation provides an alternative geometric perspective in which entanglement reflects the persistence of holomorphic correlations across scales.

Recursive contraction of state space corresponds naturally to thermodynamic processes. As degrees of freedom are compressed into stable correlations, the system moves toward states of lower informational uncertainty. In this sense renormalisation resembles a learning process, interpretable as an iterative selection in which only correlations stable under coarse-graining are retained. Structures that persist become part of the effective description of the system, while unstable fluctuations are discarded. The contraction factor $\frac{1}{\alpha}$ therefore measures how rapidly uncertainty is compressed during this learning cycle. Early recursion steps produce large gains in predictive structure, while later steps yield progressively smaller refinements as the system approaches a stable attractor.

Energy and information ultimately propagate through the universe along null trajectories carried by photons, and the concept of Twistor space was originally developed precisely to describe such light-like structures, where each twistor represents an entire null ray, encoding both its direction and its phase. Because all forms of energy eventually radiate through photon emission, entropy flow can be interpreted as the expansion of these null congruences. In this view the spreading of light cones corresponds to the propagation of informational uncertainty through spacetime.
Thermodynamic gradients therefore acquire a natural geometric interpretation. Entropy gradients correspond to curvature in the underlying informational geometry, while equilibration corresponds to the flattening of that curvature as correlations distribute themselves more evenly. The conformal-entropic correspondence therefore provides a geometric picture of how complexity emerges in recursive systems. Each recursion step performs three operations:

- contraction of uncertainty
- transport of phase correlations
- projection onto scale-invariant structure

Together these operations generate the logarithmic spirals and fractal hierarchies commonly observed at the edge of chaos. In informational terms this process resembles iterative learning. The universe continually tests possible correlations, compresses those that remain stable, and discards those that do not. Over many cycles the system converges toward attractors that encode the most persistent correlations. Complex structure therefore appears through repeated refinement of informational stability.

If recursive contraction governs the growth of correlations, several observable consequences may follow. First, entanglement entropy in many-body systems should approach saturation through scale-invariant increments consistent with Feigenbaum-type contraction. Experiments using cold atoms or superconducting qubits may provide a testing ground for this behaviour. Second, cosmological evolution should reflect the gradual accumulation of stable correlations through black hole formation. If black holes act as long-term repositories of information, their population may influence the effective thermodynamic balance of the universe. Finally, recursive contraction implies diminishing informational returns at higher levels of organisation. This may help explain why complexity grows rapidly in early evolutionary stages but slows as systems approach stable attractors.

The conformal-entropic correspondence proposes that recursive structure formation can be interpreted geometrically as a conformal contraction with phase transport in a complex informational manifold. Twistor geometry provides a natural mathematical language for describing this process because it encodes scale, orientation, and null propagation within a unified framework. Within this picture renormalisation becomes an informational recursion. Each iteration compresses uncertainty, preserves stable correlations, and produces scale-invariant structures that reappear at higher levels of organisation. Complexity therefore arises from the repeated contraction of uncertainty into memory, a process that gradually shapes the hierarchy of structure observed throughout the universe.

Nonlinearity and the Redistribution of Retained Structure

Recursive contraction alone cannot explain the emergence of structure. If each renormalisation step merely reduced amplitude uniformly, systems would shrink toward triviality, progressively losing detail without generating new forms of organisation. What makes recursive systems generative is nonlinearity, because the structure that survives each contraction is not preserved evenly. Instead, it is selectively redistributed, with some correlations sharpened, others suppressed, and phase relations reorganised into increasingly coherent patterns. It is this uneven redistribution that transforms contraction into structure and enables the formation of attractors. In nonlinear dynamics, this behaviour is captured by the Feigenbaum constant δ, which governs how structures are rescaled across successive bifurcations. While α determines the rate at which degrees of freedom are compressed, δ determines how the surviving structure is internally reorganised. If α sets the magnitude of what survives, δ sets the geometry of what survives. In the DE framework, this distinction is fundamental. Structure then emerges because the remaining correlations are reshaped into configurations that can persist under further recursion. This can be incorporated directly into the recursive dynamics. The linear contraction–rotation operator,

$$Z_{n+1} = \frac{1}{\alpha} e^{i\phi} Z_n,$$

captures the reduction of unresolved freedom and the preservation of phase. However, to account for nonlinear redistribution, the evolution must be extended to

$$Z_{n+1} = \frac{1}{\alpha} e^{i\phi} \mathcal{R}_\delta(Z_n),$$

where $\mathcal{R}_\delta$ is a nonlinear map governing the internal rescaling of the retained informational state. This operator acts selectively on the surviving correlations, concentrating those that remain stable under recursion while dispersing or suppressing those that do not. The result is that each iteration both reduces and reshapes the system, progressively concentrating structure into a smaller number of highly correlated degrees of freedom. From the DE perspective, this redistribution corresponds to a reorganisation of memory density. If Υ represents committed structure, then nonlinear rescaling increases the concentration of that structure within a shrinking subset of the informational manifold. Regions of phase alignment become increasingly dominant, while incoherent regions are eliminated. Over successive iterations, this produces a landscape in which correlations are no longer evenly distributed, but clustered into tightly coupled domains. These domains behave as unified entities because their internal degrees of freedom are no longer independent.

This then provides a natural route to attractor formation. As nonlinear redistribution proceeds, the system is driven toward configurations in which the surviving correlations reinforce one another under further contraction. These configurations define attractors in the informational manifold. Unlike trivial fixed points, these attractors retain internal structure and support continued recursion. They are therefore the stable carriers of memory across scales.
The role of δ becomes particularly clear in this context, because it governs how distances and correlations are

rescaled unevenly, it determines how quickly the system transitions from a diffuse collection of weakly related components to a tightly organised structure. When redistribution is mild, correlations remain broadly distributed and many configurations remain viable. When redistribution is strong, correlations concentrate rapidly, and the system is driven toward a small number of highly constrained states. In the latter case, the system approaches a regime in which its behaviour is effectively determined by the structure of the attractor itself.

This transition can be understood as a loss of effective independence. Although the system may retain many nominal degrees of freedom, nonlinear redistribution reduces their effective dimensionality by binding them into correlated clusters. As a result, the number of independent configurations shrinks more rapidly than would be expected from contraction alone. Predictability increases because fewer independent possibilities remain. In this sense, predictability is the consequence of lost independence. At sufficiently high levels of recursion, this process leads to a qualitative shift in behaviour, where the system no longer acts as a collection of loosely interacting components, but as a single, integrated entity governed by its attractor geometry. This is the regime in which entanglement-like behaviour emerges. Correlations become so concentrated and interdependent that they cannot be decomposed into independent parts without destroying the structure itself. Entanglement, in this view, is the limiting case of nonlinear correlation concentration under repeated renormalisation.

The combined action of α and δ therefore defines the geometry of emergence. The α-law governs the rate at which uncertainty is compressed into structure, while the δ-law governs how that structure is reorganised into increasingly coherent form. Together they ensure that recursive systems become more ordered, more correlated, and ultimately more unified. What appears at higher levels of organisation is not simply a reduced version of what came before, but a new attractor shaped by the nonlinear concentration of memory.

Renormalisation, Holography, and the Geometry of Memory

The conformal-entropic correspondence acquires a deeper structure when viewed through the combined lens of renormalisation and holography. Renormalisation describes how systems evolve under changes of scale, progressively eliminating microscopic degrees of freedom while retaining only those correlations that remain stable. Within the DE framework, this process corresponds to the compression of uncertainty into persistent informational structure. Each renormalisation step reduces accessible freedom, preserves viable correlations, and encodes them into memory.

However, renormalisation alone does not specify how these retained correlations are organised across scales. That role is governed by the nonlinear redistribution described by the Feigenbaum constant δ. While α determines how much of the system survives each contraction, δ determines how the surviving structure is reshaped. The recursion therefore does not simply compress, it reorganises, concentrating correlations into increasingly coherent domains. The result is not a uniform reduction of structure, but the progressive formation of attractors with nontrivial internal geometry.

This becomes particularly significant when interpreted through holography. In holographic theories, the additional spatial dimension of the bulk geometry behaves as a scale parameter of the boundary theory. From the DE perspective, this radial direction can be identified with recursion depth. Moving into the bulk corresponds to successive applications of the transformation

$$Z_{n+1} = \frac{1}{\alpha} e^{i\phi} \mathcal{R}_{\delta}(Z_n),$$

so that each increment in the radial coordinate represents an additional step in the compression and redistribution of information. In this interpretation, the bulk geometry becomes the accumulated record of recursive informational filtering. Each layer of the geometry encodes the correlations that survived at that level of coarse-graining, while its curvature reflects how those correlations were redistributed. The large-scale structure of spacetime therefore emerges as a geometric trace of the system's renormalisation history. This perspective also sharpens the interpretation of entanglement. In holographic models, entanglement entropy is associated with minimal surfaces connecting regions of the boundary. Within the DE framework, these surfaces can be understood as boundaries between domains of stabilised memory. They mark the regions over which correlations have been successfully preserved through recursive contraction. The area of such a surface therefore measures not only the strength of correlation, but the extent of shared informational history between subsystems. The holographic direction thus acquires a precise informational meaning, because it represents the depth to which uncertainty has been compressed and reorganised into stable structure. Renormalisation identifies which correlations persist, nonlinear redistribution shapes how they are organised, and holography encodes the result as geometry. Entropy then measures the cost of this transformation, quantifying the amount of uncertainty that has been converted into memory. Taken together, these ideas describe a single underlying process, where the universe recursively refines its own structure. Each step compresses uncertainty, redistributes correlations, and encodes the result into a geometry that reflects its history. What appears as spacetime is therefore the accumulated record of this process, a hierarchy of structures shaped by the repeated conversion of uncertainty into stable, correlated form.

Entanglement, Memory, and the Collapse of Complexity

Entanglement is often treated as if increasing it always increases richness, but that is only true up to a point. In the DE framework, entanglement has two distinct regimes. In the constructive regime, correlations bind degrees of freedom without eliminating them. Structure becomes layered and hierarchical, and complexity grows. In the constrictive regime, correlations become so global that degrees of freedom are no longer independently addressable, so that structure no longer expands, rather it starts to become increasingly compressed.

Early along the complexity axis, correlations are local and modular. Each renormalisation step reduces degrees of freedom, but in this regime the reduction is efficient, removing redundancy while preserving the independent modes required for further organisation. Quarks bind into hadrons, atoms into molecules, and molecules into living systems. Entanglement therefore compresses the system without collapsing it, reducing dimensionality while retaining the capacity for further structure. Complexity increases because reduction is selective, and not exhaustive.

As recursion proceeds, this balance begins to shift. Correlations extend across larger scales, becoming less modular and more globally coupled. The same process that once enabled structure now starts to constrain it. A turning point is reached where further entanglement no longer supports new organisation but instead reduces the number of independently addressable degrees of freedom. The system gradually loses its internal separability, transitioning from a collection of interacting components to a single, tightly bound entity.

At the limit of this process, entanglement saturates. The system can no longer sustain internal variation, and complexity collapses into substrate memory. A black hole represents this endpoint, where correlations are fully global and only a single effective macrostate remains.

This transition can be expressed quantitatively. Let the system consist of components $X_1, \dots, X_N$, and define total correlation

$$I_{\text{tot}} = \sum_i H(X_i) - H(X_1, \dots, X_N).$$

An effective number of independent degrees of freedom can then be written as

$$D_{\text{eff}} = N - \frac{I_{\text{tot}}}{\bar{H}}.$$

As entanglement increases, I_{tot} grows and D_{eff} decreases. Predictive memory can be defined as

$$I_{\text{pred}} = I(X_{<\varepsilon}; X_{>\varepsilon}),$$

and complexity follows naturally as

$$C = I_{\text{pred}} D_{\text{eff}}.$$

Entanglement, Memory, and the Collapse of Complexity

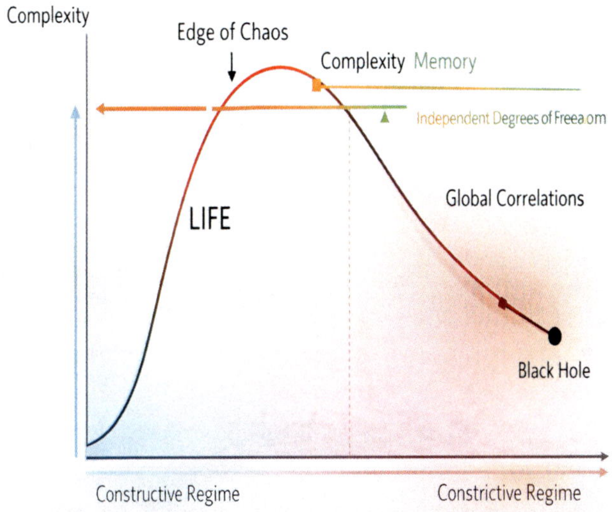

This relation captures the full behaviour of the system. Early in the recursion, entanglement increases predictive memory faster than it reduces independence, so complexity rises. Later, as correlations saturate, independence collapses and complexity falls. The peak occurs when the gain in predictive memory is exactly offset by the loss of independent degrees of freedom, and the edge of chaos corresponds precisely to this maximum. It is the point where binding and freedom are balanced, where memory persists but variability remains. Too little correlation and structure cannot stabilise, too much correlation and structure cannot vary, and only near this balance does sustained complexity exist.

In the DE framework, the edge of chaos is the only region of state space where complexity is possible. Below it, memory dissipates too readily, and above it degrees of freedom collapse too quickly. Living systems appear finely tuned because the viable region is intrinsically narrow and structurally defined. If the universe possesses a finite recursion budget, as suggested by the cosmological bound λ, then complexity must unfold within a limited hierarchy of stabilisation thresholds. The known transitions, from particles to chemistry to life, occupy only a small fraction of this depth. Most of the recursion ladder remains unexplored.

This implies that complexity is still unfolding. Just as earlier thresholds enabled new forms of organisation, higher thresholds may permit structures that are not yet visible or even conceptually accessible. The universe therefore generates structure progressively, constrained by a finite but deep recursion capacity. What we observe today may represent only the early stages of this process. The structures familiar to physics, chemistry, and biology may be the opening layers of a much larger hierarchy of organisation. Complexity grows, peaks, and collapses locally, yet across the full recursion landscape the process continues. The universe is not a finished system, but an evolving record of its own compression, redistributing uncertainty into memory across scales until its recursion budget is exhausted.

Feigenbaum Scaling and the Hubble Tension: A Renormalisation Cascade

The discrepancy in measured values of the Hubble constant is typically framed as a conflict between early- and late-universe observations. Within the standard cosmological model, this tension demands resolution through additional parameters or new physics. Within the DE framework, however, the divergence is neither anomalous nor unexpected. It reflects a deeper structural feature of reality, namely that expansion is governed by a renormalisation cascade, and that cascade follows universal scaling laws.

At the heart of this interpretation lies a well-established result from nonlinear dynamics. Systems that approach chaos through iterative bifurcation do so in a highly ordered way, characterised by the Feigenbaum constants. The most relevant of these is the spatial scaling constant, $\alpha \approx 2.5029$, which governs how attractor structures compress across successive bifurcations. This constant is universal and it appears wherever recursive thresholding and renormalisation occur, independent of the underlying substrate.

In the DE framework, this universality takes on physical meaning. Each bifurcation corresponds to a collapse event, a transition in which unresolved configurational freedom is reduced and stabilised into structure. These events are not continuous but discrete, forming a ladder of renormalisation steps. With each step, the system undergoes compression, formalised through the operator Υ, and advances along the entropic coordinate ε. The evolution of the system is therefore not smooth but structured, with scaling relationships inherited from the geometry of recursion itself. This structure is encoded in the governing relation:

$$R(\varepsilon) = -\frac{dS}{d\varepsilon}$$

which defines the rate of collapse as the gradient of entropy along the entropic coordinate. Crucially, this gradient is not uniform, because it is shaped by the discrete bifurcation structure of the system, and therefore inherits Feigenbaum scaling. The collapse rate varies therefore in accordance with a universal compression law, and the connection to cosmology follows directly. The Hubble parameter, conventionally defined as $H = \dot{a}/a$, is a kinematic measure of how rapidly distances expand. In the DE framework, this expansion is seen as a geometric projection of the same collapse dynamics. The rate at which space appears to expand reflects the rate at which entropy is being resolved into structure. Thus,

$$H(\varepsilon, \Upsilon) \sim R(\varepsilon)$$

and the observed expansion rate inherits the scaling properties of the collapse process. When Feigenbaum scaling is introduced, this relationship becomes more specific. Each renormalisation level n corresponds to a distinct structural scale, and the collapse rate transforms accordingly. To leading order, this can be expressed as:

$$H_n \sim H_0 \cdot \alpha^n$$

or its inverse, depending on normalisation. The key point is that the expansion rate is not constant across scales, but systematically rescaled at each level of the renormalisation hierarchy. This provides a natural explanation for the

Hubble tension. Observations of the early universe, such as those derived from the Planck mission, probe a regime of low renormalisation depth. Structural memory, represented by Υ, is minimal, and the system remains close to its initial symmetric recursion. The inferred value of H_0 in this regime is therefore lower. In contrast, late-universe measurements using the Hubble Space Telescope and distance ladder techniques probe a regime of significantly higher renormalisation depth. Structure has accumulated through successive collapse events, the entropic gradient has steepened, and the observed expansion rate is correspondingly higher. The ratio between these measurements is reflects the difference in renormalisation depth between the regimes being sampled:

$$\frac{H_{\text{late}}}{H_{\text{early}}} \sim \alpha^{\Delta n}$$

where Δn represents the number of effective bifurcation steps separating the two observational domains. The Hubble tension is therefore a direct manifestation of the Feigenbaum scaling embedded within the collapse geometry of the universe.

This reframing has significant implications. It removes the need to treat the discrepancy as a failure of the cosmological model or as evidence for new, unobserved components such as early dark energy. Instead, it reveals that the assumption of a single global expansion constant is itself invalid. The universe does not expand uniformly, it renormalises. Each scale carries its own effective expansion rate, determined by its position within the bifurcation hierarchy. More broadly, this establishes a deep connection between cosmology and nonlinear dynamics. The same universal constants that govern the onset of chaos in simple iterative systems also govern the large-scale structure of the universe. Expansion, structure formation, and entropy resolution are not separate processes but different expressions of a single underlying recursion. In this light, the Hubble tension is the observational imprint of a renormalising universe, one in which expansion is quantised across scales and governed by universal laws of collapse.

Every new layer of organised structure must fit into a rapidly shrinking space of possibility, forcing topology to compress toward a critical boundary where order and chaos meet. In the classic period-doubling cascade we see control parameter values $\mu_1, \mu_2, \mu_3 \ldots$ where bifurcations occur and the gaps shrink according to:

$$\mu_{n+1} - \mu_n$$

Sigma (δ) is:

$$\delta = \lim_{n\to\infty} \frac{\mu_n - \mu_{n-1}}{\mu_{n+1} - \mu_n} \approx 4.669$$

In plain terms each new structural transition happens ~4.7× closer than the previous one. Here sigma is informing us how fast the "space of stability" collapses as complexity increases. The key intuition is that every bifurcation creates a new branch and a new possible state, where each new "region" in parameter space the behaviour is stable but those regions shrink fast and exponentially with ratio ~4.66. Topology depends on how many regions exist, how they are separated, and how accessible they are. Sigma governs all three variables because it creates "crowding" of structure where each new layer of structure sits inside a much smaller interval, and yet closer to instability. So topology becomes densely packed near a critical boundary.

Therefore we see wide, simple regions initially, then rapidly compressed layers, followed by a then an almost continuous "foam" of possibilities.

This creates a boundary of accumulation, where all those shrinking intervals converge to a single point (the Feigenbaum point). That point is hence a topological attractor, so the system develops a singular boundary where infinitely many structures may accumulate:

$$\mu_n \to \mu_\infty \text{ as } n \to \infty$$

This is a classic accumulation point. However in practice real systems have limits such as noise, finite precision, quantum limits ($\hbar$), and thermal fluctuations. So in reality $n \leq n_{\max}$ and $\Delta\epsilon_n \geq \epsilon_{\min}$, and what we actually see is not literally infinity but unbounded complexity. This has three direct consequences:

A. Loss of separability

States become indistinguishable at finite resolution

B. Sensitivity explosion

Tiny changes shift between many possible structures

C. Dense attractor structure

Topology becomes "continuous-like" despite discrete origins. So as Feigenbaum scaling compresses ε-space, distinguishable states become densely packed and interactions between components become unavoidable. This drives a breakdown of factorisability, forcing the system to be described as a unified state. In classical systems this appears as complex attractor structure, and in quantum systems, the analogous condition is entanglement.

Entanglement is the quantum analogue of factorisation failure driven by recursive compression.

Therefore where early structures form easily, later structures require extreme fine-tuning. This means Υ accumulation becomes increasingly expensive in ε. The topological consequence create a very specific complexity structure. Broad, low-resolution attractors are easy to access, stable, and low Υ cost. Rapidly narrowing higher-order attractors are tightly packed, with high sensitivity and fragility. Then at the critical boundary (Feigenbaum point) we see infinite potential structure, with zero tolerance for error. Feigenbaum sigma enforces this rule by acting as a universal constraint on how quickly ε-space can be partitioned into stable Υ states, so that complexity can increase, but the room available for it collapses exponentially. Systems are forced therefore toward either simple stable structure, or critical edge behaviour. This implies there is a hard limit on scalable structure, and beyond that systems become metastable or chaotic;

$$\Delta\epsilon_{n+1} \approx \frac{\Delta\epsilon_n}{\delta_F}$$

where:

$$\delta_F \approx 4.669$$

This means each successive organisational transition has only about:

$$\frac{1}{4.669} \approx 0.214$$

of the ε-room available to the previous one. So the system becomes increasingly "more complex", yet at the same time it becomes more tightly constrained.

The Metastable Attractor

Scalable structure persists only while this contracted ε-window remains larger than the coherence threshold required for memory propagation. Once the ε-window falls below that threshold, the system may still form local structure, but it can no longer transmit that structure across scale;

$$n^* = \frac{\ln(\Delta\varepsilon_0/\varepsilon_{\text{coherence}})}{\ln \delta_F}$$

where:

- n^*= recursion depth at which scalable structure fails
- $\Delta\varepsilon_0$= initial ε-tolerance
- $\varepsilon_{\text{coherence}}$= minimum ε-width required for memory to remain coherent
- δ_F= Feigenbaum sigma/delta ≈ 4.669

After n^*:

$$n > n^*$$

This is the DE definition of metastability

$$\boxed{\text{Metastability occurs when } \frac{\Delta\epsilon_0}{\delta_F^n} < \epsilon_{\text{coherence}}}$$

A structure becomes metastable when the ε-space available to sustain it becomes smaller than the ε-space needed to preserve coherent memory.

Feigenbaum δ

The value 4.669… falls out as the only scaling factor that lets the period-doubling structure reproduce itself exactly under renormalisation. Take a simple nonlinear map (logistic map is the classic). As we tune the control parameter:

the bifurcation points μ_n accumulate. The renormalisation requirement is a transformation that "zooms in" on this cascade and returns the same structure. Operator $\mathcal{R}$:

$$\mathcal{R}[f](x) = \lambda^{-1} f(f(\lambda x))$$

A fixed point satisfies:

$$f^*(x) = \lambda^{-1} f^*(f^*(\lambda x))$$

Now linearise around the fixed point:

$$f = f^* + \varepsilon g$$

Apply the operator:

$$\mathcal{R}[f^* + \varepsilon g] \approx f^* + \varepsilon\, \mathcal{L}[g]$$

This produces an eigenvalue problem:

$$\mathcal{L}[g] = \delta\, g$$

That eigenvalue is:

$$\delta \approx 4.669$$

This limits how perturbations grow under renormalisation and δ is the only growth factor that preserves the self-similar structure under doubling and rescaling. If δ were smaller intervals would not shrink fast enough and the cascade would not converge. If δ were larger the intervals would collapse too fast and structure would not replicate across scales.

It appears therefore that in many systems because the renormalisation depends only on the presence of a single quadratic maximum, all systems demonstrate the same δ value as the intrinsic amplification factor of mismatch under recursive compression. That balance defines the edge of scalable structure, hence as the eigenvalue of the only self-consistent solution to recursive doubling with scale invariance. It is the only rate at which structure can be repeatedly doubled, compressed, and re-aligned without breaking its own topology. So new memory is created at points of non-factorisability.

Each renormalisation step asks can this still be described in the old basis? If yes no new memory accumulates, and if no new effective degree of freedom appear. That is the basis of memory formation, where components that were previously separable are forced into interaction, marking the transition from separable topology to entangled topology.

$$\boxed{\mathcal{M}_{n+1} \text{ forms when } I(A_n; B_n) > I_{\text{crit}}}$$

where:

- $A_n$and B_n are component structures at recursion level n
- $I(A_n; B_n)$is their mutual information
- I_{crit} is the threshold beyond which the system can no longer treat them as independent

Before threshold:

$$P(A_n, B_n) \approx P(A_n)P(B_n)$$

The attractor is factorisable. After threshold:

$$P(A_n, B_n) \neq P(A_n)P(B_n)$$

The attractor becomes non-factorisable. That is the birth of new memory. As $\Delta\epsilon_n$ shrinks, separable components are forced into tighter coupling. So mutual information rises across recursion:

$$I_{n+1} \sim \delta_F^{\gamma} I_n$$

where γ measures how strongly ε-compression amplifies coupling. So memory forms when:

$$\boxed{\delta_F^{\gamma n} I_0 > I_{\text{crit}}}$$

or equivalently:

$$\boxed{n^* = \frac{\ln\,(I_{\text{crit}}/I_0)}{\gamma \ln\,\delta_F}}$$

This provides us with the recursion depth at which factorisation fails;

$$\boxed{\mathcal{M}_{n+1} \Leftrightarrow P(A_n, B_n) \neq P(A_n)P(B_n)\ \text{under}\Delta\epsilon_{n+1} = \Delta\epsilon_n/\delta_F}$$

Time Dilation

Time is the rate at which new memory can be formed given available ε-space;

$$\text{time} \sim \frac{d\Upsilon}{d\varepsilon}$$

Using Feigenbaum contraction we have:

$$\Delta\varepsilon_{n+1} = \frac{\Delta\varepsilon_n}{\delta_F} \text{with } \delta_F \approx 4.669$$

So with each recursive layer less ε available, where finer distinctions are required with tighter constraints on change. To generate new Υ (memory), the system must resolve distinctions in ε crossing attractor boundaries. Then if:

$$\Delta\varepsilon \to 0$$

transitions require finer and finer precision and state changes become harder to resolve. So:

$$\frac{d\Upsilon}{d\varepsilon} \uparrow \Rightarrow \frac{d\varepsilon}{d\Upsilon} \downarrow$$

Hence more "effort" (structure change) is required per unit ε, and as ε-space contracts under Feigenbaum scaling, the system requires increasingly fine resolution to produce new distinguishable states. This reduces the effective rate at which new memory can be formed, giving rise to an apparent slowing of time. At the Feigenbaum point we see infinite bifurcations with zero ε spacing between them, so in other words infinite potential structure but with no resolvable transitions, hence explaining the dynamical freezing of time at the edge of chaos.

This aligns with known behaviour where near critical points we see a critical slowing down of time. It also explains why near horizons time dilation occurs, whilst near phase transitions long-lived states emerge. It is the same pattern repeating itself, because when the system cannot resolve change, time appears to slow:

$$\text{effective time rate} \;\propto\; \Delta\epsilon_n$$

So under Feigenbaum scaling:

$$t_{n+1} \sim \frac{t_n}{\delta_F}$$

each deeper level evolves more slowly relative to its structure.

Under recursive renormalisation, Feigenbaum scaling compresses ε-space by a factor of approximately 4.669 at each bifurcation layer. Because the formation of new memory requires resolvable distinctions in ε, this contraction progressively limits the system's capacity to generate new states. As a result, the effective rate of change decreases, and the system exhibits a form of dynamical slowing. Time, in this framework, is an emergent measure of the rate at which ε can be resolved into new Υ. As ε-space approaches zero at the accumulation point, the system approaches a state of maximal structural density and minimal dynamical evolution.

The rate of observable change is limited by the resolution of ε-space. So after n recursions:

$$\Delta\epsilon_n = \frac{\Delta\epsilon_0}{\delta_F^n}$$

If define time operationally:

$$t_n \propto \frac{1}{\text{rate of resolvable transitions}}$$

If transitions require resolving ε, then:

$$t_n \propto \frac{1}{\Delta\epsilon_n}$$

So:

$$\boxed{t_n \sim \delta_F^n}$$

Therefore as structure deepens ε-space contracts exponentially and time expands exponentially, so that deep structure evolves slowly and shallow structure evolves quickly. Then if memory forms when factorisation fails and that happens at recursion depth n^*, we can say:

$$t_{n^*} \sim \delta_F^{n^*}$$

and hence the birth of new memory is associated with a sharp increase in effective timescale, where discrete jumps in ϒ correspond to dilation in time. From here we can predict;

A. Hierarchical time scales

- fast microdynamics
- slow macrodynamics
- locked by δ

B. Critical freezing

- near accumulation point → $t \to \infty$

C. Discrete emergence

- new organisational layers come with time dilation

Feigenbaum scaling provides a direct bridge between renormalisation and the emergence of time. With each recursive bifurcation, the available ε-space contracts by a factor ≈ 4.669. Because the formation of new distinguishable states requires finite ε-resolution, this contraction reduces the rate at which new states can be resolved. Defining time operationally as the inverse rate of resolvable transitions yields $t_n \sim \delta_F^n$. Time therefore emerges as the dual of ε under renormalisation, as ε contracts, time dilates. This establishes a quantitative link between hierarchical structure, memory formation, and the emergence of slow dynamics at large scales, because Feigenbaum scaling sets the rate at which reality can change.

General Relativity

g_{00} is the geometric encoding of Feigenbaum-compressed ε-resolution

In GR, time slows because the metric component g_{00} changes. So we can say;

$$\Delta\varepsilon_n \downarrow \Rightarrow \text{resolvable transitions slow} \Rightarrow d\tau \downarrow \Rightarrow g_{00} \downarrow$$

Under Feigenbaum renormalisation $\Delta\varepsilon_n = \frac{\Delta\varepsilon_0}{\delta_F^n}$ and effective transition time expands as $t_n \sim \delta_F^n$, as so;

$$\text{rate of change}_n \sim \frac{1}{t_n} \sim \delta_F^{-n}$$

In weak-field GR:

$$d\tau = \sqrt{g_{00}}\, dt$$

and approximately:

$$g_{00} \approx 1 + \frac{2\Phi}{c^2}$$

where Φ is gravitational potential.

Near stronger gravity proper time slows, clocks tick more slowly relative to distant observers, and g_{00} decreases. If gravity is the geometric bookkeeping of constrained history then:

$$\sqrt{g_{00}}$$

is the observable residue of how much ε-resolution remains available for local change. That gives:

$$\boxed{\sqrt{g_{00}} \sim \frac{\Delta\epsilon_n}{\Delta\epsilon_0}}$$

Using Feigenbaum scaling:

$$\boxed{\sqrt{g_{00}} \sim \delta_F^{-n}}$$

Therefore:

$$\boxed{g_{00} \sim \delta_F^{-2n}}$$

This means that each deeper renormalisation layer compresses ε-space by δ_F and reduces local proper-time rate by δ_F, which maps geometrically into g_{00}. So the DE-to-GR bridge becomes:

$$\boxed{d\tau_n \sim \delta_F^{-n} dt}$$

The Planck Scale

At the Planck scale, ε-resolution cannot shrink indefinitely. There is a minimum distinguishable action step:

$$\Delta\Sigma_{\text{min}} = \hbar$$

or equivalently:

$$\oint d\Sigma = 2\pi n\hbar$$

That means ε-compression eventually hits a quantisation floor;

$$\Delta\varepsilon_n \not\to 0$$

The maximum recursion depth is therefore:

$$\boxed{n_P = \frac{\ln\left(\Delta\varepsilon_0/\varepsilon_P\right)}{\ln \delta_F}}$$

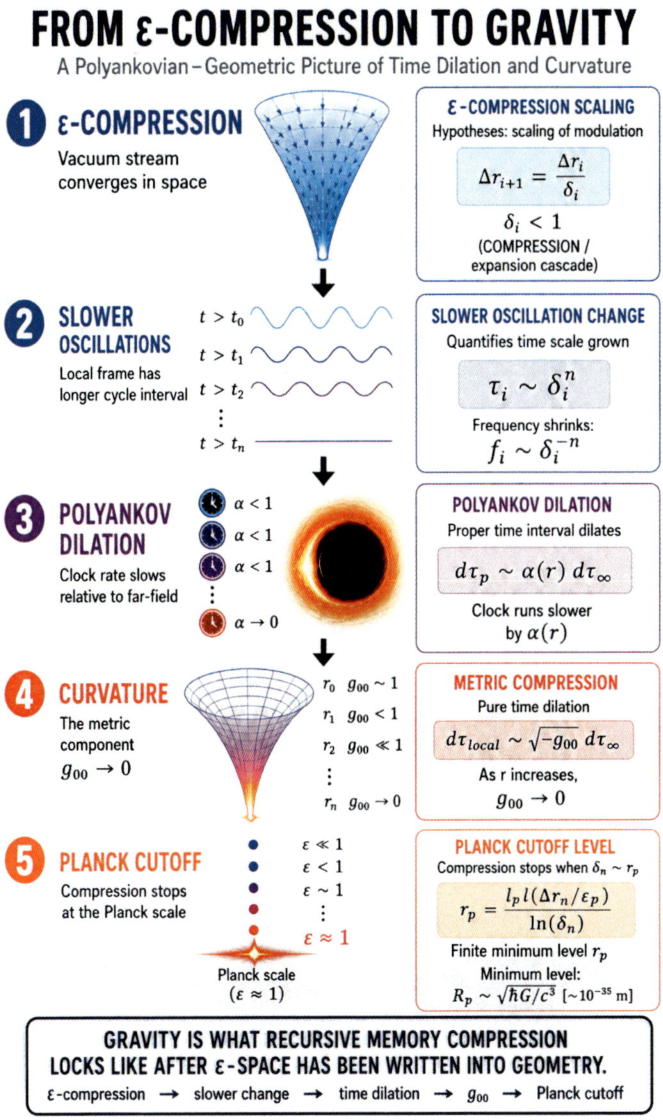

Feigenbaum renormalisation provides a direct bridge between recursive structure formation and relativistic time. At each bifurcation layer, the available ε-resolution contracts by $\delta_F \approx 4.669$, so the rate of resolvable transition falls as δ_F^{-n}. In general relativity, local proper time is encoded by $d\tau = \sqrt{g_{00}}dt$. The DE interpretation is that $\sqrt{g_{00}}$ represents the geometric encoding of remaining ε-resolution. Thus $\sqrt{g_{00}} \sim \delta_F^{-n}$, and $g_{00} \sim \delta_F^{-2n}$. Gravity is therefore the large-scale geometric representation of recursive ε-compression, so that as history becomes more deeply constrained, local change slows, proper time dilates, and the metric records the cost of sustaining accumulated memory. At the Planck scale, this contraction cannot continue indefinitely because ε-resolution reaches the minimum action threshold $\Delta\Sigma_{\text{min}} = \hbar$, creating a natural cutoff to recursive compression.

This framework proposes therefore that spacetime and quantum behaviour emerge from a single underlying process, namely recursive ε-compression governed by Feigenbaum scaling. Gravity appears as the geometric encoding of accumulated memory, while quantum discreteness arises from a lower bound on ε-resolution.

Weak-Field Gravity

In the weak-field limit, gravity may be interpreted as the geometric expression of recursive memory compression. As non-factorisable structure accumulates, local ε-resolution contracts, reducing the rate of resolvable change. Since proper time measures the local rate of change, $\frac{d\tau}{dt}$ becomes proportional to the remaining ε-resolution. The metric component g_{00} therefore encodes local ε-compression. If memory density is denoted ρ_Υ, then $\sqrt{g_{00}} \sim e^{-\kappa\rho_\Upsilon}$, giving $\Phi = -\kappa c^2 \rho_\Upsilon$ in the weak-field approximation. Spatial gradients in accumulated memory density generate gravitational acceleration, while the Poisson equation emerges when matter density acts as the source term for curvature in ρ_Υ. Feigenbaum scaling supplies the discrete renormalisation law, so each recursive layer contracts ε-space by δ_F, giving $\sqrt{g_{00}} \sim \delta_F^{-n}$. Thus gravity is the large-scale metric representation of recursively compressed memory. Starting with the DE premise;

$$\text{memory density} = \rho_\Upsilon$$

where ρ_Υ means accumulated non-factorisable structure per unit volume. In ordinary gravity, the weak-field metric is:

$$g_{00} \approx 1 + \frac{2\Phi}{c^2}$$

where Φ is the gravitational potential. Our DE reinterpretation is:

$$\Phi \sim -c^2\,\mathcal{C}(\rho_\Upsilon)$$

where $\mathcal{C}$ is the local constraint cost generated by accumulated memory. So;

$$\boxed{\text{gravity} = \text{the spatial gradient of memory-compression cost}}$$

Let local ε-resolution fall as memory density rises:

$$\Delta\varepsilon(x) = \Delta\varepsilon_0 e^{-\kappa\rho_\Upsilon(x)}$$

where κ measures how strongly accumulated memory constrains future resolution. Then proper time follows remaining ε -resolution:

$$\frac{d\tau}{dt} \sim \frac{\Delta\varepsilon(x)}{\Delta\varepsilon_0}$$

so:

$$\frac{d\tau}{dt} = e^{-\kappa\rho_{\Upsilon}(x)}$$

In GR:

$$\frac{d\tau}{dt} = \sqrt{g_{00}}$$

Therefore:

$$\boxed{\sqrt{g_{00}} = e^{-\kappa\rho_{\Upsilon}(x)}}$$

For weak fields:

$$g_{00} \approx 1 - 2\kappa\rho_{\Upsilon}(x)$$

Compare with:

$$g_{00} \approx 1 + \frac{2\Phi}{c^2}$$

So:

$$\boxed{\Phi(x) = -\kappa c^2 \rho_{\Upsilon}(x)}$$

Gravity appears when memory density varies spatially:

$$\vec{g} = -\nabla\Phi$$

Substitute the DE potential:

$$\vec{g} = \kappa c^2 \nabla \rho_\Upsilon$$

So:

$$\boxed{\vec{g} \sim \nabla \rho_\Upsilon}$$

Hence bodies fall because they move along gradients of accumulated constraint. Newtonian gravity obeys:

$$\nabla^2 \Phi = 4\pi G \rho$$

In DE form:

$$\Phi = -\kappa c^2 \rho_\Upsilon$$

so:

$$\nabla^2 \Phi = -\kappa c^2 \nabla^2 \rho_\Upsilon$$

To recover Newtonian gravity, require:

$$-\kappa c^2 \nabla^2 \rho_\Upsilon = 4\pi G \rho$$

or:

$$\boxed{\nabla^2 \rho_\Upsilon = -\frac{4\pi G}{\kappa c^2} \rho}$$

This is the core weak-field DE gravity equation. Matter generates a curvature in memory-density space, and that curvature becomes gravitational potential. At recursion depth n:

$$\Delta \varepsilon_n = \frac{\Delta \varepsilon_0}{\delta_F^n}$$

and:

$$\frac{d\tau_n}{dt} \sim \delta_F^{-n}$$

So if memory density corresponds to effective recursion depth:

$$n(x) \sim \rho_\Upsilon(x)$$

then:

$$\sqrt{g_{00}} \sim \delta_F^{-n(x)}$$

Taking logs:

$$\ln \sqrt{g_{00}} = -n(x) \ln \delta_F$$

For weak fields:

$$\sqrt{g_{00}} \approx 1 + \frac{\Phi}{c^2}$$

so:

$$\boxed{\frac{\Phi}{c^2} \approx -n(x) \ln \delta_F}$$

From renormalisation to weak-field gravity;

$$\boxed{\sqrt{g_{00}} \sim \frac{\Delta\varepsilon}{\Delta\varepsilon_0} \sim \delta_F^{-n} \sim e^{-\kappa \rho_\Upsilon}}$$

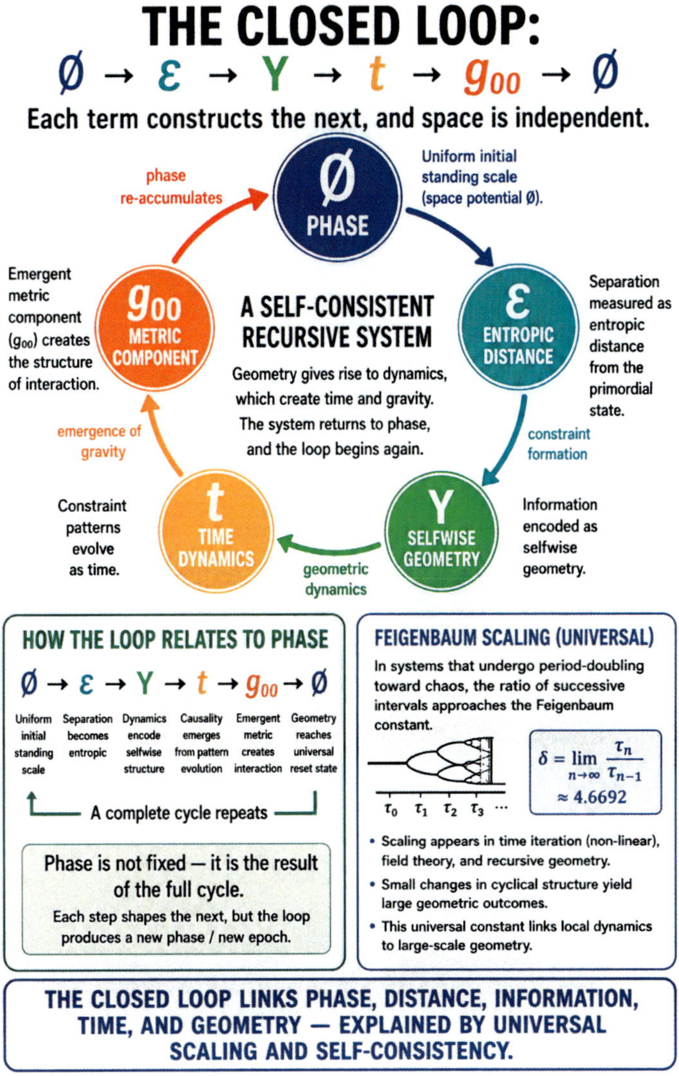

Dark Matter

Dark-matter-like behaviour may emerge if non-factorisable constraint, generated by baryonic structure under recursive renormalisation, forms an extended, non-dissipative field whose spatial profile approximates $1/r^2$. In this case, the additional curvature attributed to dark matter would instead arise from the geometric effects of accumulated constraint. Any derivation must satisfy the following conditions;

Condition 1: It must carry effective energy density

$$\rho_{\text{constraint}} > 0$$

Condition 2: It must extend spatially

Dark matter halos scale according to:

$$\rho(r) \sim \frac{1}{r^2}$$

So our constraint field must naturally produce:

$$\rho_{\Upsilon}(r) \sim \frac{1}{r^2}$$

Condition 3: It must not radiate away

Ordinary energy disperses. Therefore our essential constraint must be stable, non-dissipative, and long-lived.

So far our framework already produces renormalised structure across scales, non-factorisable memory, and persistence of constraint. Thus we can say;

$$\boxed{\text{constraint is a stored, non-radiative form of energy}}$$

Dark-matter-like behaviour may emerge therefore if non-factorisable constraint, generated by baryonic structure under recursive renormalisation, forms an extended, non-dissipative field whose spatial profile approximates $1/r^2$. In this case, the additional curvature attributed to dark matter would instead arise from the geometric effects of accumulated constraint:

$$\nabla^2 \rho_\Upsilon = -\alpha \rho_{\text{baryonic}}$$

Hence:

$$\rho_\Upsilon(r) \sim \frac{M}{r}$$

Then gravitational potential:

$$\Phi \sim -\rho_\Upsilon \sim -\frac{M}{r}$$

$$\boxed{\text{renormalisation must redistribute constraint into a } 1/r^2 \text{ profile}}$$

If;

$$\Phi \sim \log r$$

this implies:

$$\rho_\Upsilon \sim \frac{1}{r^2}$$

Feigenbaum scaling implies structure repeats across scales and constraint propagates hierarchically;

$$\rho_\Upsilon(r) = \sum_n \delta_F^{-n} f(r/\lambda^n)$$

This can then generate scale-free tails and effective power laws. Where a purely symmetric constraint field would tend to cancel or smooth out, chirality prevents this cancellation. Recursive constraint biases, and this preserves orientation across scale. So the can be viewed as a chiral memory field.

$$\rho_\Upsilon(r, \theta) = \rho_0(r) + \rho_\chi(r, \theta)$$

where:

- $\rho_0(r)$= radial constraint density

- $\rho_{\chi}(r, \theta)$= chiral/asymmetric correction

Because galaxies are rotating, handed, angular-momentum-bearing systems, a chiral constraint field naturally couples to rotation. Not all entropy becomes observable structure and unemergent entropy supplies the hidden reservoir. Some entropy remains unresolved, non-radiative, non-factorised, and below emergence threshold and we can consider this as an unemergent entropic burden;

$$S_{\text{total}} = S_{\text{emergent}} + S_{\text{unemergent}}$$

Thus dark-matter-like behaviour may arise from the second term:

$$S_{\text{unemergent}} \rightarrow \rho_{\Upsilon,\text{latent}} \rightarrow \text{extra curvature}$$

hence:

$$\boxed{T_{\mu\nu}^{\text{eff}} = T_{\mu\nu}^{\text{baryon}} + T_{\mu\nu}^{\Upsilon} + T_{\mu\nu}^{\chi} + T_{\mu\nu}^{S_u}}$$

where:

- $T_{\mu\nu}^{\Upsilon}$= accumulated memory constraint
- $T_{\mu\nu}^{\chi}$= chiral orientation / rotational bias
- $T_{\mu\nu}^{S_u}$= unemergent entropy contribution

Dark matter is the gravitational signature of unresolved, chiral, non-factorised entropy.

Feigenbaum scaling supplies the radial hierarchy:

$$\rho_{\Upsilon}(r) \sim \sum_{n} \delta_F^{-n} f(r/\lambda^n)$$

This can generate a scale-free halo. Chirality modifies the field:

$$\rho_{\Upsilon}(r) \rightarrow \rho_{\Upsilon}(r) + \chi\,\Omega(r)$$

where $\Omega(r)$ encodes rotational handedness. Unemergent entropy can be expressed as:

$$\rho_{\text{dark}} \sim \rho_{\Upsilon} + \rho_{\chi} + \rho_{S_u}$$

In the DE interpretation, dark-matter-like behaviour therefore arises from the interaction of three hidden structural terms: accumulated non-factorisable memory, chiral orientation, and unemergent entropy. Feigenbaum scaling distributes constraint across radius, chirality prevents cancellation and couples the field to rotation, while unemergent entropy supplies a non-radiative reservoir of unresolved structure. The resulting field contributes to curvature without appearing as ordinary luminous matter.

Across recursive renormalisation, the contraction of ε-space governed by Feigenbaum scaling forces a compensatory accumulation of non-factorisable structure. The ratio $\mathcal{J} = \frac{\Delta\Upsilon}{\Delta\varepsilon}$ remains approximately invariant across scales and defines the cost of converting uncertainty into memory. This invariant appears as a coupling constant linking source density to curvature in physics as the gravitational constant, and in complex systems as constraint-driven dynamics. Gravity is therefore the large-scale manifestation of a scale-invariant constraint ratio. Across all these steps, one quantity keeps reappearing:

$$\boxed{\mathcal{J} \equiv \frac{\Delta\Upsilon}{\Delta\varepsilon}}$$

where:

- $\Delta\Upsilon$: newly committed (non-factorisable) structure
- $\Delta\varepsilon$: resolution available to distinguish it

- And $\mathcal{J}$ is the cost of turning uncertainty into memory.

That is the quantity that survives renormalisation.

However without compensating mechanism, $\mathcal{J}$ would blow up with n. The only way the system remains stable is:

$$\Delta\Upsilon_n \sim \delta_F^n \, \Delta\varepsilon_n$$

so that:

$$\mathcal{J}_n \approx \text{constant}$$

And;

$$\boxed{G_{\text{DE}} \sim \mathcal{J} = \frac{\Delta\Upsilon}{\Delta\varepsilon}}$$

Where G is literally resistance to future change and the invariant ratio of memory formation to resolution, i.e. inertia / gravitation in generalised form.

At each stage non-factorisable structure accumulates as Υ, and recursive compression drives ε downward, forcing the system to continually re-express itself within shrinking resolution. In a purely symmetric system, these contributions would cancel as constraint would average out, and no persistent field would emerge across scales. But real systems are not symmetric, because chirality breaks that cancellation. It introduces orientation, a directional bias that prevents recursive structure from simply negating itself. Constraint acquires handedness which persists and accumulates across scales, and this asymmetry allows memory to propagate rather than dissolve.

At the same time, not all structure becomes visible. A portion of Υ remains below the threshold of emergence. This unemergent entropy does not radiate, decohere, nor dissipate. Instead it is retained as latent constraint, contributing to the system without appearing as explicit structure. It is the hidden reservoir that stabilises the field. The result is that the key invariant of the system survives compression:

$$\mathcal{J} = \frac{\Delta(\Upsilon_{\text{emergent}} + \Upsilon_{\text{latent}})}{\Delta\varepsilon}$$

This ratio is the cost of converting uncertainty into memory. It is scale invariant and robust under renormalisation, so it persists as the thread connecting regimes that otherwise appear unrelated. At the microscale it governs fast dynamics and weak constraint, at the mesoscale it shapes attractors and biological organisation, and at the macroscale it manifests as curvature and gravity. Scaling therefore creates complexity while preserving constraint, and through chirality and unemergent entropy, that constraint becomes directional, stable, and cumulative. What emerges is a field that is a persistent imprint of recursive compression that survives across scales. This is why the same ratio appears everywhere where scaling creates structure, because it preserves coupling across scales.

Appendix A: Formalisation of the Informational State Variable Z

This appendix provides a formal definition of the complex informational variable Z introduced in the Conformal–Entropic Correspondence (CEC), and specifies its evolution under recursive stabilisation, including the role of nonlinear redistribution governed by the Feigenbaum constant δ.

A1. Definition of the Informational State

Let the local informational state of a system be represented by a complex variable

$$Z \in \mathcal{M}_I \cong \mathbb{C}, Z = \rho\, e^{i\theta}.$$

Here, $\mathcal{M}_I$ denotes a local complex informational manifold. The components of Z have the following interpretation:

- **Amplitude ρ**: measures the effective magnitude of unresolved configurational freedom.
- **Phase θ**: encodes the relational ordering of correlations within the system.

Within the DE framework, the amplitude is defined as a function of the entropic coordinate ε and the structural memory Υ:

$$\rho = f(\varepsilon, \Upsilon), \frac{\partial f}{\partial \varepsilon} > 0, \frac{\partial f}{\partial \Upsilon} < 0.$$

A minimal representation is given by

$$Z = \frac{\varepsilon}{\Upsilon} e^{i\theta},$$

so that Z increases with unresolved uncertainty and decreases as stable structure accumulates.

A2. Recursive Contraction and Redistribution Operator

A single stabilisation step is represented by a composite operator $\mathcal{R}_\delta$ acting on Z:

$$\mathcal{R}_\delta(Z) = \frac{1}{\alpha} e^{i\phi}\, \mathcal{D}_\delta(Z),$$

where:

- $\alpha \approx 2.503$ is the Feigenbaum contraction constant,
- ϕ is the phase increment per recursion step,
- $\mathcal{D}_\delta$ is a nonlinear redistribution operator governed by the Feigenbaum constant δ.

The iterative dynamics are therefore

$$Z_{n+1} = \frac{1}{\alpha} e^{i\phi}\, \mathcal{D}_\delta(Z_n).$$

In contrast to the purely linear case, the operator $\mathcal{D}_\delta$ reshapes the retained structure after each contraction, concentrating viable correlations and suppressing unstable modes.

A3. Spiral Convergence with Nonlinear Redistribution

In the absence of nonlinear effects, the solution reduces to

$$Z_n = Z_0\, \alpha^{-n} e^{in\phi},$$

yielding

$$\mid Z_n \mid = \mid Z_0 \mid\ \alpha^{-n}, \arg(Z_n) = \arg(Z_0) + n\phi.$$

This corresponds to a logarithmic spiral in $\mathcal{M}_I$.

With nonlinear redistribution, the trajectory remains contractive and rotational but becomes geometrically structured. The operator $\mathcal{D}_\delta$ introduces:

- uneven contraction of phase-space regions,
- clustering of correlations,
- sharpening of attractor structure.

Thus:

- **Amplitude decay** corresponds to compression of uncertainty,
- **Phase transport** preserves relational correlations,
- **δ-redistribution** reshapes the internal geometry of the surviving structure.

Without δ, contraction leads to uniform decay, but with δ, contraction generates structured attractors.

A4. Relation to Entropy, Memory, and Redistribution

Substituting

$$Z_n = \left(\frac{\varepsilon_n}{\Upsilon_n}\right) e^{i\theta_n},$$

the recursion rule implies

$$\frac{\varepsilon_{n+1}}{\Upsilon_{n+1}} = \frac{1}{\alpha}\frac{\varepsilon_n}{\Upsilon_n}, \theta_{n+1} = \theta_n + \phi,$$

with an additional redistribution of structure

$$\Upsilon_{n+1} = \mathcal{D}_\delta(\Upsilon_n).$$

Each recursion step therefore:

- reduces unresolved uncertainty relative to committed structure by $1/\alpha$,
- preserves correlation geometry through phase continuity,
- redistributes memory density according to δ.

In this sense:

- α governs how much structure survives,
- δ governs how that structure is reorganised.

A5. Fixed Points, Attractors, and Scale Invariance

In the linear representation, the only strict fixed point satisfies

$$Z_* = \frac{1}{\alpha} e^{i\phi} Z_*,$$

giving

$$Z_* = 0.$$

However, physical systems exhibit non-trivial stable structures. This is resolved by recognising equivalence under scale transformations:

$$Z \sim \alpha Z,$$

or by working in projective space

$$[Z] \in \mathbb{CP}^1.$$

With nonlinear redistribution, the attractor is thus a structured invariant set determined by repeated application of $\mathcal{D}_\delta$. Stability corresponds to invariance of correlation geometry rather than absolute magnitude.

A6. Conformal and Twistor Extension

For systems where orientation, scale, and null propagation are coupled, Z can be generalised to a twistor-like object

$$Z^A \in \mathbb{T},$$

with recursion

$$Z^A_{n+1} = \frac{1}{\alpha} e^{i\phi} \, M^A{}_B \, \mathcal{D}_\delta(Z^B_n),$$

where $M^A{}_B$ is a conformal transformation preserving projective relations among null directions. In this formulation:

- $\frac{1}{\alpha}$ contracts informational amplitude,
- $e^{i\phi}$ transports phase,
- $\mathcal{D}_\delta$ redistributes internal structure,
- M preserves conformal correlation geometry.

A7. Continuous Limit

In the continuum approximation,

$$\frac{dZ}{dn} = (-\ln \, \alpha + i\phi) \, Z + \mathcal{N}_\delta(Z),$$

where $\mathcal{N}_\delta$ represents the nonlinear contribution induced by δ. The linear solution remains

$$Z(n) = Z(0)\exp\left[(-\ln\ \alpha + i\phi)n\right],$$

while deviations from this trajectory encode the emergence of structured attractors through nonlinear redistribution.

A8. Interpretation

The variable Z provides a compact representation of informational state:

- Its **modulus** encodes residual uncertainty,
- Its **phase** encodes persistent correlation structure,
- Its **nonlinear evolution** encodes the redistribution of memory.

The recursion operator therefore captures the full mechanism of structure formation:

$$\begin{aligned}\text{uncertainty contraction} &+ \text{phase preservation}\\ &+ \text{nonlinear redistribution}\\ &\longrightarrow\ \text{stable, scale-invariant structure.}\end{aligned}$$

Within this formalism:

- renormalisation corresponds to recursive contraction,
- δ governs the geometry of retained memory,
- attractors emerge through repeated redistribution of correlations.

Thus, renormalisation, learning, and thermodynamic equilibration can be understood as manifestations of a single process, with the progressive compression of uncertainty into structured memory, shaped not only by how much is retained, but by how that retained structure is reorganised across scales.

Appendix B: Feigenbaum Scaling and the Renormalisation of the Hubble Parameter

This appendix refines the formal relationship between entropy collapse, Feigenbaum scaling, and the observed scale-dependence of the Hubble parameter within the Deductive Entropy (DE) framework. In particular, it resolves an important subtlety, that while renormalisation proceeds through discrete bifurcation events, cosmological observables emerge from statistical ensembles and therefore appear continuous.

B1. Entropic Collapse and Expansion

We begin with the core DE relation:

$$R(\varepsilon) = -\frac{dS}{d\varepsilon}$$

where:

- ε is the entropic coordinate
- $S(\varepsilon)$ is unresolved configurational entropy
- $R(\varepsilon)$ is the collapse rate

In DE, spatial expansion is the geometric projection of collapse. The Hubble parameter,

$$H = \frac{\dot{a}}{a},$$

is therefore identified with the collapse rate:

$$H(\varepsilon, \Upsilon) \propto R(\varepsilon)$$

where Υ represents accumulated structural memory.

B2. Discrete Renormalisation

Collapse proceeds through discrete bifurcation events. Each event corresponds to a renormalisation step indexed by:

$$n \in \mathbb{Z}$$

At each level:

- structure is compressed
- degrees of freedom are reduced
- memory Υ increases

These steps form a renormalisation hierarchy rather than a continuous trajectory.

B3. Feigenbaum Scaling

For recursive bifurcation systems, the Feigenbaum constant

$$\alpha \approx 2.5029$$

governs the scaling of structure between successive levels. In DE, this defines how collapse rates transform across renormalisation depth:

$$R_n = R_0\, \alpha^n$$

and therefore:

$$H_n \propto R_n \Rightarrow H_n = H_0\, \alpha^n$$

This relation holds at the level of individual, fully resolved renormalisation states.

B4. Ensemble Nature of Cosmological Observables

The universe does not exist at a single renormalisation level. At any given epoch, it contains a distribution of structures spanning multiple depths. Define:

$P(n)$
$= \text{distribution of structures across renormalisation levels}$

The observed Hubble parameter is therefore an ensemble average:

$$H_{\text{obs}} = \int H(n)\, P(n)\, dn$$

Substituting the Feigenbaum scaling:

$$H_{\text{obs}} = H_0 \int \alpha^n P(n)\, dn$$

This introduces an effective renormalisation depth:

$$\bar{n} = \int n\, P(n)\, dn$$

such that:

$$H_{\text{obs}} \approx H_0\, \alpha^{\bar{n}}$$

To leading order, the observable behaves as though it arises from a continuous depth parameter $\bar{n} \in \mathbb{R}$, even though the underlying dynamics are discrete.

B5. Early- and Late-Universe Measurements

Let:

- $P_{\text{early}}(n)$: narrow distribution at low renormalisation depth (e.g. CMB measurements from the Planck mission)
- $P_{\text{late}}(n)$: broader distribution extending to higher depths (e.g. distance ladder measurements using the Hubble Space Telescope)

Then:

$$H_{\text{early}} \approx H_0\, \alpha^{\bar{n}_{\text{early}}}, H_{\text{late}} \approx H_0\, \alpha^{\bar{n}_{\text{late}}}$$

and:

$$\frac{H_{\text{late}}}{H_{\text{early}}} = \alpha^{\Delta \bar{n}}$$

where:

$$\Delta \bar{n} = \bar{n}_{\text{late}} - \bar{n}_{\text{early}}$$

B6. Interpretation of Fractional Scaling

The observed ratio:

$$\frac{H_{\text{late}}}{H_{\text{early}}} \approx \frac{73}{67} \approx 1.09$$

yields:

$$\Delta \bar{n} \approx \frac{\log{(1.09)}}{\log{(\alpha)}} \approx 0.10$$

This reflects a shift in the ensemble-averaged renormalisation depth. In other words:

- Individual collapse events remain discrete
- Observables reflect statistical weighting across many completed levels

A small change in the distribution $P(n)$ produces a fractional shift in $\bar{n}$, and therefore a smooth change in H.

B7. Continuous Effective Formulation

The Hubble parameter can therefore be written as:

$$H(\varepsilon, \Upsilon) = H_*\, \alpha^{\bar{n}(\varepsilon, \Upsilon)}$$

where $\bar{n}$ is a continuous function encoding the cumulative distribution of collapse states. This reconciles:

- discrete renormalisation at the micro level
- continuous observables at the macro level

B8. Consequences

This formulation yields several testable implications:

1. **Scale dependence**
   H varies systematically with observational scale via shifts in $\bar{n}$

2. **Structure dependence**
   Regions with greater structural memory
   Υ contribute disproportionately to higher Hvalues.

3. **Smooth transitions**
   Intermediate probes (e.g. BAO) should yield intermediate values through gradual shifts in $P(n)$

4. **Non-arbitrary divergence**
   Differences in measured H_0 follow structured scaling rather than random offset.

B9. Summary

Renormalisation in DE is fundamentally discrete, governed by universal Feigenbaum scaling. However, cosmological observables arise from ensembles of structures distributed across these discrete levels. This produces an effective continuous renormalisation depth, allowing small, fractional differences in observed expansion rates. The Hubble tension is therefore the natural outcome of measuring a renormalising universe through ensemble-averaged observables.

Toy model: DE weak-field gravity from ε-compression

Consider a one-dimensional system in which accumulated non-factorisable memory is represented by a density field $\rho_\Upsilon(x)$. The central DE assumption is that local ε-resolution contracts as memory accumulates: $\Delta\varepsilon(x) = \Delta\varepsilon_0 e^{-\kappa\rho_\Upsilon(x)}$. Since proper time measures the local rate of resolvable change, $d\tau/dt = \Delta\varepsilon/\Delta\varepsilon_0$, giving $\sqrt{g_{00}} = e^{-\kappa\rho_\Upsilon}$. In the weak-field limit this becomes $g_{00} \approx 1 - 2\kappa\rho_\Upsilon$, which matches $g_{00} \approx 1 + \frac{2\Phi}{c^2}$ if $\Phi = -\kappa c^2\rho_\Upsilon$. Gravitational acceleration is therefore the gradient of memory-compression cost: $a = \kappa c^2\nabla\rho_\Upsilon$. If memory depth is discrete and Feigenbaum-scaled, then $\sqrt{g_{00}} = \delta_F^{-n(x)}$, linking gravitational time dilation directly to recursive ε-compression.

Final toy-model chain:

$$\boxed{\rho_\Upsilon(x) \to \Delta\epsilon(x) \to d\tau/dt \to g_{00} \to \Phi \to a}$$

Assume a 1D space x, with a central mass-like region that carries accumulated memory density:

$$\rho_\Upsilon(x)$$

Let ε -resolution fall where memory density is high:

$$\Delta\varepsilon(x) = \Delta\varepsilon_0 e^{-\kappa\rho_\Upsilon(x)}$$

Define proper-time rate as remaining ε -resolution:

$$\frac{d\tau}{dt} = \frac{\Delta\varepsilon(x)}{\Delta\varepsilon_0} = e^{-\kappa\rho_\Upsilon(x)}$$

But in weak-field GR:

$$\frac{d\tau}{dt} = \sqrt{g_{00}}$$

So:

$$\boxed{\sqrt{g_{00}(x)} = e^{-\kappa \rho_\Upsilon(x)}}$$

For weak memory density:

$$e^{-\kappa \rho_\Upsilon} \approx 1 - \kappa \rho_\Upsilon$$

so:

$$g_{00} \approx 1 - 2\kappa \rho_\Upsilon$$

Compare with:

$$g_{00} \approx 1 + \frac{2\Phi}{c^2}$$

Therefore:

$$\boxed{\Phi(x) = -\kappa c^2 \rho_\Upsilon(x)}$$

That is the toy gravitational potential.

Motion rule

A test particle moves down the gradient of the potential:

$$a(x) = -\frac{d\Phi}{dx}$$

Substitute:

$$\Phi(x) = -\kappa c^2 \rho_\Upsilon(x)$$

Then:

$$\boxed{a(x) = \kappa c^2 \frac{d\rho_\Upsilon}{dx}}$$

So in the toy model acceleration is generated by the gradient of memory density. A body falls because it moves toward increasing accumulated constraint.

Add Feigenbaum scaling

Let memory depth be discrete:

$$n(x) = \lambda \rho_\Upsilon(x)$$

Feigenbaum compression gives:

$$\frac{\Delta\varepsilon_n}{\Delta\varepsilon_0} = \delta_F^{-n}$$

So:

$$\frac{d\tau}{dt} = \delta_F^{-n(x)}$$

Therefore:

$$\boxed{\sqrt{g_{00}(x)} = \delta_F^{-n(x)}}$$

and:

$$\boxed{g_{00}(x) = \delta_F^{-2n(x)}}$$

For weak fields:

$$\ln\sqrt{g_{00}} = -n(x)\ln\,\delta_F$$

Since:

$$\sqrt{g_{00}} \approx 1 + \frac{\Phi}{c^2}$$

we obtain:

$$\boxed{\frac{\Phi(x)}{c^2} \approx -n(x)\ln\,\delta_F}$$

So Feigenbaum gives the discrete scaling law behind gravitational time dilation.

Minimal trajectory equation

Define a DE action:

$$\Sigma[x] = \int \left[\frac{1}{2}m\dot{x}^2 - m\kappa c^2\rho_\Upsilon(x)\right]dt$$

This is just ordinary kinetic cost minus memory-compression potential. The Euler-Lagrange equation gives:

$$m\ddot{x} = m\kappa c^2\frac{d\rho_\Upsilon}{dx}$$

so:

$$\boxed{\ddot{x} = \kappa c^2\frac{d\rho_\Upsilon}{dx}}$$

Which matches the earlier acceleration rule.

Scaling does two things simultaneously

Feigenbaum scaling:

$$\Delta\varepsilon_{n+1} = \frac{\Delta\varepsilon_n}{\delta_F}$$

forces:

- contraction of distinguishability (ε)
- accumulation of non-factorisable structure (Υ)

So scaling is geometric and constraint-building. However constraint is not uniform because if every region had the same ε-compression, then time slows everywhere equally with no motion or forces, and hence no gravity;

$$\nabla\rho_{\Upsilon} \neq 0$$

Non-uniform scaling → gradients

When different regions have different recursion depth $n(x)$:

$$\sqrt{g_{00}(x)} \sim \delta_F^{-n(x)}$$

So:

- deeper recursion → more compression → slower time
- shallower recursion → less compression → faster time

This creates a time gradient

Time gradient → motion

Systems evolve along paths that minimise constraint cost. This gives:

$$a \sim \nabla \rho_\Upsilon$$

which is gravitational acceleration. Scaling therefore creates a hierarchy of ε-compression and gravity emerges when that hierarchy becomes spatially non-uniform.

Newtonian Gravity

To recover Newtonian gravity, the DE framework distinguishes between memory density ρ_Υ and mass-energy density ρ. Mass-energy does not equal memory; rather, it acts as a source term that induces curvature in the memory field. This is expressed by the relation $\nabla^2 \rho_\Upsilon = -\alpha\rho$, where $\alpha = 4\pi G/(\kappa c^2)$. The gravitational potential is then given by $\Phi = -\kappa c^2 \rho_\Upsilon$, which yields $\nabla^2 \Phi = 4\pi G\rho$, recovering Poisson's equation. For a point mass, this produces $\Phi = -\frac{GM}{r}$ exactly. Gravity therefore emerges as motion along gradients in a field of accumulated constraint, whose structure is governed by recursive ε-compression and Feigenbaum scaling. Newtonian limit:

$$\nabla^2 \Phi = 4\pi G\rho$$

From our toy model:

$$\Phi = -\kappa c^2 \rho_\Upsilon$$

So:

$$\nabla^2 \Phi = -\kappa c^2 \nabla^2 \rho_\Upsilon$$

To match Newton:

$$-\kappa c^2 \nabla^2 \rho_\Upsilon = 4\pi G\rho$$

Define:

$$\boxed{\nabla^2 \rho_\Upsilon = -\alpha\rho}$$

where: $\alpha = \frac{4\pi G}{\kappa c^2}$

So memory density is the source of matter density, where;

- ρ= local energy/mass
- ρ_Υ= accumulated constraint field
- matter creates spatial curvature in ρ_Υ

Solving the toy case

Let:

$$\rho(x) = M\delta(x)$$

Then:

$$\nabla^2 \rho_\Upsilon = -\alpha M \delta(x)$$

Solution:

$$\rho_\Upsilon(r) \sim \frac{\alpha M}{4\pi r}$$

Substitute into:

$$\Phi = -\kappa c^2 \rho_\Upsilon$$

gives:

$$\Phi(r) = -\kappa c^2 \cdot \frac{\alpha M}{4\pi r}$$

Insert α:

$$\Phi(r) = -\frac{GM}{r}$$

falls out cleanly with no tuning beyond constants.

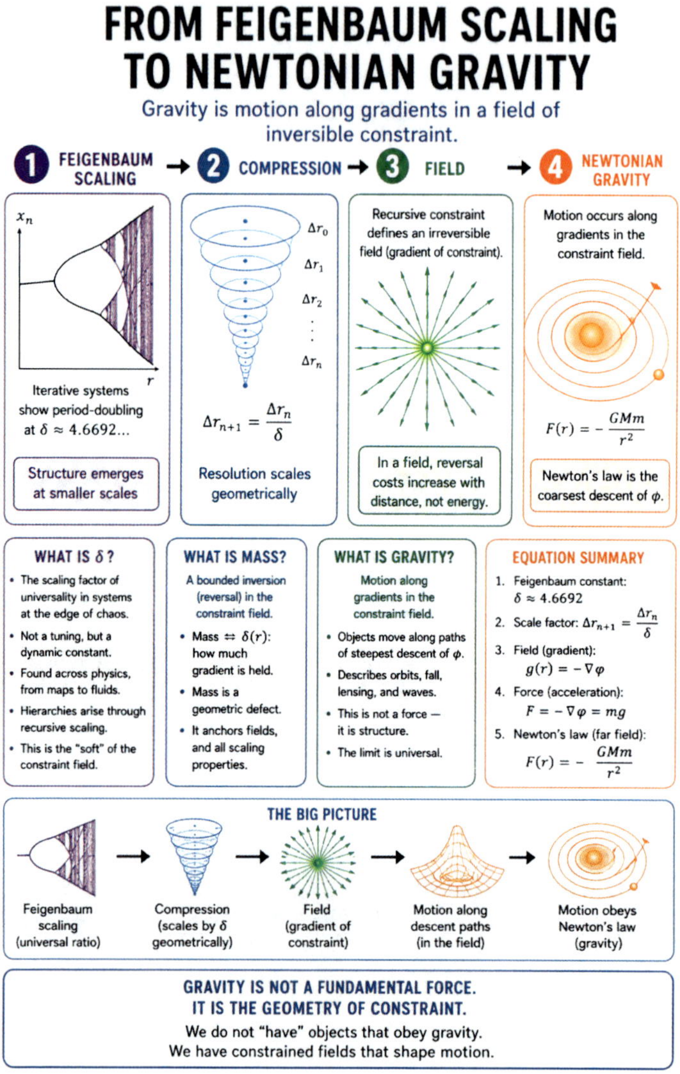

The scalar memory-field model may be extended naturally into tensor form by treating spacetime curvature as the coarse-grained representation of non-uniform recursive constraint. In the weak-field limit, accumulated memory density ρ_Υ modifies the time-time component of the metric g_{00}, through its effect on local ε-resolution. In full tensor language, this suggests a memory-curvature relation of the form $\mathcal{G}_{\mu\nu}^{(\Upsilon)} = \chi T_{\mu\nu}$, where $T_{\mu\nu}$ acts as the source term for non-uniform memory compression. When coarse-grained, $\mathcal{G}_{\mu\nu}^{(\Upsilon)}$ becomes the usual Einstein curvature tensor, recovering general relativity in the classical limit. Gravity therefore appears as the tensor geometry of recursively compressed memory. We can now close the loop:

$$\rho_\Upsilon(x) \sim n(x)$$

where:

- $n=$ local recursion depth

and:

$$\sqrt{g_{00}} \sim \delta_F^{-n(x)}$$

So:

$$\boxed{\text{mass-energy determines how deeply recursion proceeds locally}}$$

which determines:

- ε-compression
- time dilation
- curvature

Tensor upgrade

Scalar toy model:

$$\nabla^2 \rho_\Upsilon = -\alpha\rho$$

Tensor version:

$$\boxed{\mathcal{G}_{\mu\nu}^{(\Upsilon)} = \chi T_{\mu\nu}}$$

where:

- $T_{\mu\nu}$= mass-energy source
- $\mathcal{G}^{(\Upsilon)}_{\mu\nu}$= curvature of the memory/constraint field
- χ= coupling constant, chosen so the weak-field limit recovers $8\pi G/c^4$

This mirrors Einstein's equation:

$$G_{\mu\nu} = \frac{8\pi G}{c^4} T_{\mu\nu}$$

but interprets curvature as geometry emerging from non-uniform recursive compression. More precisely:

$$T_{\mu\nu} \rightarrow \text{non-uniform memory compression} \rightarrow g_{\mu\nu} \rightarrow \text{curvature}$$

So spacetime curvature is the coarse-grained tensor representation of non-uniform Υ-density. Define a memory metric:

$$g^{(\Upsilon)}_{\mu\nu} = \eta_{\mu\nu} + h^{(\Upsilon)}_{\mu\nu}$$

with:

$$h^{(\Upsilon)}_{00} \approx -\frac{2\kappa\rho_\Upsilon}{c^2}$$

Then in weak field:

$$g_{00} \approx 1 + \frac{2\Phi}{c^2}$$

and because:

$$\Phi = -\kappa c^2 \rho_\Upsilon$$

the scalar toy model drops straight into the tensor metric.

$$\boxed{G_{\mu\nu} \text{ is the macroscopic shadow of } \nabla_\mu \nabla_\nu \Upsilon}$$

Chapter 11: The Geometry of Life

The origin of life and the subsequent emergence of multicellular organisms are often treated as contingent historical events driven by molecular innovation or genetic novelty. This chapter presents an alternative account grounded in DE, in which these transitions arise as necessary threshold events on a recursion-driven complexity axis. Within this framework, time is not the organising variable. Instead, systems evolve through recursive encounters with uncertainty, indexed by ε, while realised structure is expressed along the ϒ axis. At specific thresholds, triadic closure becomes possible, producing new attractor states in the $\Upsilon - \varepsilon$ plane that are capable of retaining memory against decoherence. Life and multicellularity are shown to be successive ratchets on this axis, each defined by the closure of a distinct SU(3)-like triad, the emergence of a higher-order attractor, and the suppression of characteristic false attractors. This account reframes biology as a geometric consequence of memory formation under uncertainty, rather than as a sequence of accidental molecular discoveries.

1. Introduction: From Dynamics to Recursion

Most physical and biological theories describe change along a temporal axis, but DE replaces time with recursion. Systems do not progress because time passes, but because they repeatedly encounter uncertainty and either resolve or fail to resolve it. Each recursive encounter either flattens entropy, erasing structure, or curves entropy, retaining constraint. Complexity therefore emerges as the result of renormalisation across recursion. When a system resolves uncertainty in a way that reduces future entropy cost, that resolution is compressed and stabilised, and memory is the persistence of this compression. Where no renormalisation occurs structure is transient, but where renormalisation is active, constraint accumulates and new attractors appear in the $\Upsilon - \varepsilon$ plane. The central claim of this chapter is that major biological transitions correspond to discrete thresholds at which renormalisation becomes self-sustaining under triadic closure.

2. Chemistry: Exploration Without Renormalisation

At the chemical level, matter responds to gradients. Energy differences drive reactions, products form and disperse, and under repeated conditions similar outcomes may recur. Complex chemistry can generate cycles, oscillations, and autocatalytic motifs that appear ordered. However, these systems do not retain successful configurations. Each cycle begins again from effectively the same starting point. There is no mechanism by which prior resolution reduces future uncertainty, so in DE terms, recursion occurs without renormalisation.

While chemistry therefore explores the $\Upsilon - \varepsilon$ plane, it does not stabilise it. Entropy changes locally but remains effectively flat across cycles. No curvature is retained, no memory accumulates, and no true attractor forms. Apparent persistence at this level reflects repetition without compression. False attractors may arise when longevity is mistaken for memory. Reaction cycles and replicating molecules can endure, but they do not improve their own stability across recursion. Without renormalisation, they remain fragile and reversible.

3. The First Threshold: The Emergence of Life

Life begins when a chemical system crosses a critical threshold at which memory becomes self-sustaining. This transition is defined by the convergence of triadic closure and renormalisation. The minimal life configuration consists of three constraints operating simultaneously:

- continuous energy throughput
- internal constraint retention
- selective boundary control

Individually, each is insufficient. Together, they form a closed loop in which perturbations are absorbed and corrected rather than amplified. Triadic closure establishes the possibility of stability, while renormalisation makes that stability persistent. Each cycle of perturbation and correction now reinforces the same solution. Successful configurations are compressed, reducing the accessible microstate space while preserving function. At this point curvature along ε is no longer transient, it is retained and deepened across recursion. Thus self-sustaining attractor emerges in the $\Upsilon - \varepsilon$ plane, and that is what we call life.

The system no longer depends on externally imposed structure. Following perturbation, it actively restores its configuration, returning to the same region of state space through internally generated dynamics, and therefore maintaining the preservation of organisation. The governing relationship

$$R(\varepsilon) = -\frac{dS}{d\varepsilon}$$

becomes internally operative. The system shapes its own entropy gradients, and those gradients stabilise its structure, so memory emerges as the cumulative result of repeated renormalisation. The defining phenotype of life is therefore homeostasis. Material components turn over continuously, yet organisation persists because it is continuously re-stabilised, and below this threshold, systems remain open. Viruses, prions, and template replicators may exhibit stability or replication, but they do not renormalise their own dynamics. Their apparent memory accumulation is borrowed from external systems, however they do not complete the recursion, and life begins only where renormalised closure becomes self-sustaining.

4. Single-Cell Life: Closed Memory Systems

A unicellular organism is a fully renormalised system. All constraints required for persistence are internal, and each recursive cycle reinforces the system's stability. Within the $\Upsilon - \varepsilon$ plane, the cell occupies a deep attractor basin. Many microstates are possible along Υ, but only a narrow subset is retained because ε-curvature is continuously reinforced. Perturbations are absorbed, corrected, and incorporated into the system's ongoing stability. However, renormalisation has a cost, so that as internal complexity increases, the burden of maintaining constraint grows. Each additional layer of structure requires further correction, and the system approaches a limit at which further internal renormalisation becomes inefficient. This creates pressure for a higher-order solution.

5. The Second Threshold: Multicellular Life

Multicellular life emerges when renormalisation is redistributed across a collective. Memory is no longer confined within a single cell, but shared across many. The multicellular triad consists of:

- intercellular signalling
- adhesion
- spatial patterning

When these constraints close, renormalisation occurs at the level of the organism. The attractor shifts upward in the $\Upsilon - \varepsilon$ plane. Individual cells no longer maintain independent stability, instead their behaviour is stabilised by the collective. Differentiation reflects restriction to sub-basins within a higher-order attractor. Cells specialise because the organism's renormalised structure constrains their permissible states. False attractors at this level include colonies and tumours, because while these systems exhibit coordination, they lack fully integrated renormalisation, so they remain partially open.

6. Development, Regeneration, and Stability

Development reflects progressive renormalisation of form. Early states are flexible, with shallow curvature. Later states are canalised, as repeated stabilisation deepens ε-curvature. Multiple developmental pathways converge because the attractor exists independently of specific trajectories. Regeneration follows naturally, so when structure is damaged, remaining components re-enter the same renormalised basin and restore form. Memory is therefore geometric, encoded in the shape of the attractor itself. Disease then emerges when renormalisation weakens. Loss of curvature allows regression to earlier attractors, explaining why pathological states often resemble simpler or ancestral configurations.

7. False Attractors and the Illusion of Progress

At every threshold partial closure produces metastable basins. These systems occupy locally stable regions of the $\Upsilon - \varepsilon$ plane and exhibit some features of the transition, but they do not achieve full renormalised stability. Their persistence is conditional. In each case, elements of the triad are present, but the closure is incomplete. As a result, renormalisation does not stabilise the system across recursion. These metastable basins can persist for extended periods and may appear indistinguishable from fully realised attractors under limited observation. However, they remain vulnerable to perturbation because ε-curvature is not retained, so that disturbances are not fully absorbed and therefore accumulate, eventually displacing the system from the basin. DE predicts these states as inevitable. Whenever partial closure occurs, a shallow basin forms in the $\Upsilon - \varepsilon$ landscape. These basins capture trajectories transiently, giving the appearance of stability, but only when renormalisation deepens and stabilises the basin across recursive cycles does a true transition occur. Therefore metastable basins are pre-threshold structures, because they signal proximity to transition, but do not survive it unless renormalisation completes the closure.

8. The Ratchet Principle

Each major transition corresponds to a step in which renormalisation becomes self-sustaining at a higher level of organisation. Once established, these states cannot be reversed without catastrophic loss of structure. The sequence is lawful:

- chemistry explores without renormalisation
- life renormalises locally
- multicellular systems renormalise collectively

Further transitions extend this pattern, externalising memory while increasing fragility. Complexity increases but by the progressive deepening of renormalised attractors, so that when complexity saturates, further accumulation becomes unstable. At this boundary, systems reorganise how uncertainty is managed. Growth ceases to be structural and becomes dynamical. Stability is maintained not by adding components, but by improving renormalisation. Life emerges at this limit, as the solution that allows persistence when complexity can no longer expand.

The diagram that follows explores this regime. It shows how, at the edge of chaos, systems do not collapse into disorder nor freeze into rigidity, but settle into stable, recurring patterns. These attractors are the signature of life at peak complexity, where systems that survive not by growing more complex, but by learning how to move within constraint. This chapter examines life as the natural continuation of complexity's ascent, what happens when the ratchet can no longer turn, and persistence must be achieved through dynamics rather than accumulation.

Renormalised Memory and the Life Threshold

The minimal attractor state that instigates life is a closed-loop constraint system with a renormalising constraint, in which successful resolutions of uncertainty are compressed and retained across ε-recursion. Three conditions must be simultaneously satisfied:

- energetic dissipation with retention
- autocatalytic closure
- state-dependent response

These define triadic closure. However, life begins when closure becomes renormalised, such that each recursive cycle stabilises and deepens the same solution. When this occurs, the system enters a stable basin, with a renormalised attractor in the $\Upsilon - \varepsilon$ plane, and one that can resist perturbation despite continuous material turnover.

1. System state and entropy

The system is defined over realised structure and its internal generative model:

$$x(\varepsilon) \in \mathbb{R}^n, m(\varepsilon)$$
$$S(\varepsilon) = -\int p(x, \varepsilon) \ln p(x, \varepsilon)\, dx$$

Entropy is indexed by ε so that progression occurs through recursive encounters with uncertainty. Renormalisation enters here implicitly, so that if the system compresses $p(x, \varepsilon)$ across recursion entropy is reshaped.

2. Entropy curvature as memory

$$\kappa(\varepsilon) \equiv -\frac{\partial^2 S}{\partial \varepsilon^2}$$

- $\kappa = 0$: exploration without retention
- $\kappa > 0$: constraint persists across recursion

Curvature is the result of renormalisation. Each successful resolution reduces future uncertainty and is retained geometrically. Memory is therefore the cumulative effect of renormalised entropy curvature

3. Open-system constraint

$$\Delta S_{\text{out}}(\varepsilon) > \Delta S_{\text{int}}(\varepsilon)$$

Local order must be paid for by external dissipation. Without this, renormalisation cannot accumulate, and any internally generated curvature is erased. This condition ensures that renormalisation is thermodynamically admissible.

4. Renormalised memory recursion (governing equation)

The core DE law becomes explicitly renormalised:

$$M_{n+1} = \mathcal{R}(M_n) = M_n e^{-\Lambda_n} + \Delta\kappa_n$$

where:

- $\mathcal{R}(\cdot)$ is the renormalisation operator across ε-recursion
- $e^{-\Lambda_n}$ is the survival of prior structure
- $\Delta\kappa_n$ is newly generated curvature

This equation is compressive, so that only curvature that reduces future entropy cost survives. Renormalisation therefore acts as a filter on admissible structure.

5. Life-instigation threshold

Life emerges when renormalisation becomes self-sustaining:

$$M_{n+1} \geq M_n \quad \Leftrightarrow \quad \Delta\kappa_n \geq M_n(1 - e^{-\Lambda_n})$$

Below this threshold curvature is transient, so renormalisation fails and systems revert to exploration. Above this threshold however curvature accumulates and renormalisation deepens the basin, so memory becomes persistent. This is the point at which chemistry becomes life.

6. Renormalised attractor criterion

$$\lambda_{\perp}^{(\Upsilon-\varepsilon)}(A_{\text{life}}) < 0$$

A true attractor is one in which perturbations decay under ε-recursion. Renormalisation ensures this condition, so that perturbations are not just resisted, rather they are actively corrected and compressed into the attractor structure, which is where metastable basins fail. They may exhibit local stability, but without renormalisation, perturbations accumulate and eventually eject the system.

7. Salience-weighted renormalisation

$$M_{n+1} = M_n e^{-\Lambda_n} + w_n \Delta\kappa_{\text{raw},n}, w_n \in [0,1]$$

Renormalisation is selective.

- $w_n \approx 0$: curvature ignored → metastable basin
- $w_n \approx 1$: curvature retained → ratchet

This introduces directionality into evolution and enables higher-order systems to emerge.

8. Multicellular renormalisation

$$M_{\text{organism}} = \sum_i w_i M_i + M_{\text{inter}}$$

Renormalisation shifts scale from intracellular, to intercellular and then to organismal Memory becomes hierarchical as the attractor moves to a higher-dimensional $\Upsilon - \varepsilon$ manifold.

Everything else from metabolism, replication, genetics, selection, multicellularity, occurs downstream of this condition. As complexity ratchets upward, the system gains expressive power by adding degrees of freedom, yet this very success carries the seeds of its limitation. Increasing interconnection binds components together, and entanglement progressively narrows the space of independent action. At the point of maximal complexity, further accumulation ceases to be adaptive as freedom is no longer expanded but compressed, and the system faces a choice between rigidity and collapse. Life arises at this boundary, where persistence can no longer be achieved by adding structure, only by reorganising dynamics. What emerges is a regime in which stability is maintained through feedback, memory, and constrained motion within well-defined attractors. The focus therefore shifts from how complexity grows to how it is sustained once growth is no longer possible, setting the stage for life as a phenomenon that exists within chaos.

Multicellular Life: Renormalisation Beyond the Cell

Multicellular life represents a second, irreversible ratchet on the DE memory axis. The transition is not defined by a shift in how memory is stabilised. In unicellular systems, renormalisation is local, such that each recursive cycle reinforces internal ε-curvature within a single boundary. As complexity increases, the cost of maintaining this curvature rises, and the system approaches a saturation limit beyond which additional internal structure destabilises rather than stabilises. The multicellular transition occurs when total retained curvature exceeds what a single-cell attractor can sustain:

$$\kappa_{\text{total}} > \kappa_{\text{single-cell,max}}$$

At this threshold, renormalisation can no longer be contained within one agent. The system resolves this by factorising curvature across interacting units, shifting from local to distributed memory:

$$M_{\text{organism}} = \sum_i w_i M_i + M_{\text{inter}}$$

Here, M_i are cellular memory functions, M_{inter} is shared curvature encoded in signalling, adhesion, and spatial geometry, and w_i are stability weights. Memory is no longer localised but stabilised collectively, and the organism becomes the dominant attractor in the $\Upsilon - \varepsilon$ plane, constraining individual cells more strongly than their own internal dynamics. Differentiation follows as a consequence of this redistribution. Cells no longer retain full ε-access, but occupy restricted sub-basins of the organismal attractor:

$$\kappa_i(\varepsilon) \subset \kappa_{\text{organism}}(\varepsilon)$$

Stem cells retain broader ε-access with shallow curvature, whereas differentiated cells exhibit deeper curvature and reduced flexibility. Stability increases as freedom is constrained, and development reflects progressive renormalisation of form. Early states are weakly constrained, but repeated recursive stabilisation deepens curvature, canalising trajectories toward a shared attractor:

$$\exists\, A_{\text{body}} \subset \Upsilon \times \mathbb{R}_\varepsilon \quad \text{such that} \quad \Psi_i(\varepsilon) \to A_{\text{body}} \quad \forall i$$

Form is therefore not encoded locally, but emerges from the geometry of the attractor. Regeneration follows naturally, as remaining components re-enter the same basin and restore structure. The stability condition for multicellular life is:

$$\frac{\partial M_{\text{inter}}}{\partial \varepsilon} > \sum_i \Lambda_i \, M_i$$

Shared memory must grow faster than individual decoherence. When this condition fails, the system loses coherence. Cancer represents collapse of shared curvature:

$$M_{\text{inter}} \downarrow \quad \Rightarrow \quad M_i \text{ re-dominates}$$

Cells revert toward unicellular attractors, no longer constrained by organismal memory. In this sense, cancer is regressive, reflecting loss of higher-order renormalisation. Multicellular life can therefore be defined as the state in which memory is externalised across agents, forming a higher-order attractor that stabilises identity, behaviour, and function through shared constraint. A cell in isolation acts on local history, whilst within an organism, it is continuously stabilised by the system it inhabits. This transition is irreversible because memory depth increases and is redistributed rather than replaced. Once renormalisation operates at the collective level, the system cannot return to purely local stability without structural collapse.

SU(3)-Like Closure as the Minimal Renormalisation Structure

According to DE memory progresses through renormalisation, the compression and retention of entropy curvature across ε-recursion. For renormalisation to stabilise, perturbations must be converted into restoring structure rather than dissipated. This requires a minimal level of internal constraint. Pairwise interactions can generate correlation, but they do not support stable renormalisation. Perturbations propagate without sufficient constraint to reproject the system onto a stable manifold. As a result, curvature cannot be reliably retained.

The minimal structure that permits stable renormalisation is a triadic constraint system. With three mutually interacting modes, perturbations in one component are counteracted by the remaining two, allowing deviations to be corrected rather than amplified. This establishes transverse stability in the $\Upsilon - \varepsilon$ plane. Algebraically, this corresponds to the first non-abelian closure. In SU(3), constraint generators do not commute, and their interactions produce corrective modes. This non-commutativity is the formal expression of renormalisation pressure, as perturbations are transformed into retained curvature. The onset of life corresponds to the first ε-recursive phase-lock of such a system. Let three interacting constraint modes be represented by amplitudes ψ_1, ψ_2, ψ_3, with triadic phase:

$$\Phi = \arg(\psi_1 \psi_2 \psi_3)$$

and coherence:

$$T(\varepsilon) = | \langle e^{i\Phi} \rangle |$$

Life emerges when:

$$T(\varepsilon) > T_c$$

At this threshold, renormalisation becomes self-stabilising. Curvature is no longer transient but accumulates across recursion, forming a persistent attractor. The lowest-cost symmetric closure corresponds to a three-cycle:

$$\Phi^* = \pm \frac{2\pi}{3}$$

This phase minimises mismatch per ε-step and therefore minimises curvature loss. Once established, it acts as a stable renormalisation kernel across scales. In this framework, triads do not generate memory directly, but they provide the minimal structure through which renormalisation can operate. Hence that's why life begins when this structure locks, allowing curvature to persist and deepen across recursion.

At every emergence threshold, the transition occurs when a triadic constraint structure is not only formed, but stabilised through renormalisation, producing a new ε-attractor. False attractors arise when closure is incomplete or renormalisation fails, allowing only transient or metastable basins to form.

Why metastable basins appear at thresholds

At any emergence threshold, the system is under pressure to resolve growing uncertainty. Two elements of a triad often align first, so this partial closure produces a metastable basin in the $\Upsilon - \varepsilon$ plane. The system lingers there because it reduces uncertainty relative to the background. Locally, the basin appears stable, organised, and persistent. Yet one escape direction remains unconstrained, so stability is conditional rather than intrinsic. In DE terms, curvature may be generated but is not renormalised across ε-recursion. The memory function rises transiently, then plateaus or decays. Realised trajectories in Υ appear ordered, but stability is not sustained across recursion.

A metastable basin lacks transverse closure in the $\Upsilon - \varepsilon$ plane. Perturbations orthogonal to the active constraints are not corrected, but accumulate across recursion. This produces apparent stability along some Υ directions while allowing expansion along at least one $\Upsilon - \varepsilon$ escape direction. Each recursive step therefore leaks curvature, and the system survives only within a narrow environmental range. Attractors compete through renormalisation efficiency, not speed, size, or abundance. Let:

- κ_f: curvature generated by a metastable basin
- κ_t: curvature generated by a true attractor
- Λ: shared decoherence load

Then:

- metastable basin: $\kappa_f \lesssim \Lambda$
- true attractor: $\kappa_t > \Lambda$

Both may coexist temporarily. However, only the true attractor renormalises curvature across recursion. The metastable basin generates structure locally but fails to stabilise it globally. Metastable basins typically form earlier because they require fewer coordinated constraints. They populate the $\Upsilon - \varepsilon$ landscape first and may dominate phenomenologically. This is expected as the system explores partial closure before complete renormalisation is achieved.

Out-competition mechanism

a) Renormalisation advantage

True attractors deepen their basin with each ε-step. Metastable basins do not. As recursion proceeds, trajectories remain confined to the true basin with increasing probability, while metastable trajectories are eventually ejected.

b) Perturbation routing
Perturbations that displace systems from metastable basins are increasingly captured by the deeper true basin. Reverse transitions are rare.

c) Resource capture
True attractors tolerate wider environmental variation because renormalised structure stabilises internal dynamics. Metastable basins remain constrained to narrow conditions and are progressively excluded.

Metastable basins do not usually decay gradually. Because renormalisation fails, curvature does not compound, and eventually a perturbation exceeds the system's limited correction capacity, and the trajectory exits abruptly. This produces extinction-like dynamics. False solutions appear across all thresholds from prebiotic cycles collapse with minor perturbation, viral lineages depend on host constraints, tumours either regress or destabilise, and ideologies fragment under sufficient reality pressure. A critical result in DE is that metastable basins cannot be incrementally repaired. Completing the missing constraint requires renormalisation across recursion, not local modification. The system must exit the basin to discover a stable attractor as metastable basins cannot ratchet. They do not increase retained curvature across ε-recursion and therefore represent evolutionary cul-de-sacs, regardless of short-term success or abundance. There is no continuous path between them and true attractors.

Selection is therefore not time-based, but recursion-based. Systems that renormalise curvature persist as stable regions in the $\Upsilon - \varepsilon$ plane. Systems that do not are eliminated as their trajectories escape shallow basins. Metastable basins reduce uncertainty locally in Υ but amplify it across ε. True attractors reduce uncertainty recursively across all accessible ϒ–ε directions. Once a renormalised attractor exists, out-competition is inevitable.

Ratcheting is Impossible for False Attractors

Ratcheting requires net renormalised memory gain across ε-recursion. A system ratchets only when curvature generated at step ε_n is retained and compounded at ε_{n+1}. Metastable basins fail this condition structurally. They may generate curvature locally, but they do not renormalise it. As a result, the memory function saturates or decays, and ε-depth does not increase. The system revisits the same basin without extending its accessible state space.

The defining consequence is that no new ε-layer becomes reachable. True thresholds expand the dimensionality of the $\Upsilon - \varepsilon$ landscape. Metastable basins operate entirely within an existing projection of that space. They can explore, optimise, and persist, but they cannot deepen. This is the DE definition of an evolutionary cul-de-sac.

Metastable basins often display high replication rates, rapid diversification, and strong short-term optimisation. They may dominate locally in ϒ while remaining dependent on higher-order systems. Their apparent success reflects efficiency within a fixed ε-layer, and not progression beyond it. Selection does not rescue these systems. Selection operates on persistence within an attractor, not on the creation of new attractors. Transitions between ε-layers require renormalised closure, and no incremental adjustment within a metastable basin can supply the missing constraint. The system must exit the basin entirely, and the transition is discrete. Canonical examples include:

- **Viruses**: optimise replication without acquiring metabolism
- **Prions**: replicate structure without regulation
- **Colonies**: coordinate without organism-level memory
- **Cancer**: reverts toward unicellular autonomy
- **Ideologies**: reinforce internally without external correction

Each is locally successful and globally sterile. Only systems that renormalise curvature across recursion can ratchet. Metastable basins recycle curvature without compounding it, whereas a ratchet advances when new structure is stabilised. Metastable systems repeatedly traverse the same structure without ever extending it. All such systems are evolutionary cul-de-sacs, regardless of longevity or dominance.

Gravity-like behaviour in biology

Gravity appears in physics when mass-energy induces curvature in spacetime. In biological systems, high-density, non-factorisable structure induces curvature in an abstract constraint space. Although no physical gravitational field is generated, the resulting dynamics are formally equivalent: systems evolve along gradients of constraint, producing attractor-driven behaviour analogous to motion in a gravitational potential. From our chain in the previous chapter:

$$T_{\mu\nu} \to \rho_{\Upsilon} \to g_{\mu\nu} \to \nabla\rho_{\Upsilon}$$

In physics mass-energy creates real spacetime curvature,

whereas in biology structure and information create effective constraint fields. The mathematics takes the same form but operates at a different scale. Biological systems: accumulate very high Υ(non-factorisable structure) and organise it non-uniformly to maintain gradients (energy, information, morphology);

$$\nabla\rho_{\Upsilon} \neq 0$$

That produces directed behaviour and stability basins with attractor dynamics;

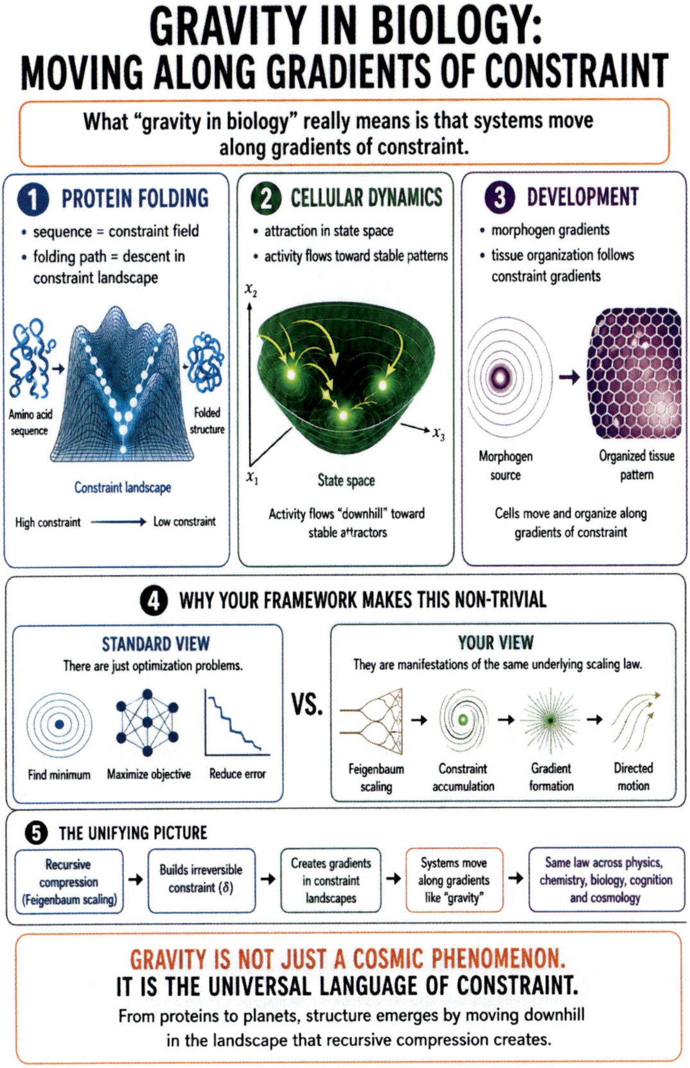

These are all manifestations of the same underlying scaling law, where Feigenbaum scaling leads to constraint accumulation, gradient formation and ultimately directed motion. This is not to "gravity exists in biology", but rather that biological systems instantiate gravity-like dynamics because they generate internal field of recursive constraint.

Therefore any system that accumulates non-factorisable structure under recursive compression will exhibit gradient-like dynamics. In physics this produces gravity, while in biology it leads to organisation, behaviour, function.

This explains why optimisation landscapes (e.g. Waddington) look "gravitational", because attractors behave like sink-wells and systems converge without explicit "force laws". They are emergent consequences of constraint geometry. Using three biological examples as scale bridges.

1. Protein folding

$$\text{amino acid sequence} \rightarrow \rho_\Upsilon \rightarrow \text{folding basin}$$

The sequence creates a constraint field. The protein falls through a shaped landscape toward a low-entropy basin.

$$\nabla \rho_\Upsilon \rightarrow \text{folding trajectory}$$

2. Neural attractors

$$\text{prior memory} \rightarrow \rho_\Upsilon \rightarrow \text{perceptual attractor}$$

The brain resolves uncertainty by moving toward high-probability attractor states. In psychosis or ASD, the constraint field may become too rigid, too shallow, or misweighted. Perception therefore maybe thought of as motion through a memory-curved inference field.

3. Cancer evolution

$$\text{mutation} + \text{inflammation} + \text{tissue stress} \rightarrow \rho_\Upsilon \rightarrow \text{malignant attractor}$$

Cancer is a shift in the tissue constraint field until malignant behaviour becomes the easier trajectory, so cancer emerges when the physiological attractor becomes the lowest-cost basin. Therefore biology recreates gravity's mathematics because living systems are dense fields of accumulated constraint. Proteins fold, brains infer, and cancers evolve by moving through landscapes shaped by non-factorisable memory. These are gravity-like in form, because systems follow gradients of constraint toward attractor basins. In physics, the field is spacetime curvature. In biology the field is functional constraint, and the same renormalisation logic applies across both. We can now develop a formulation which introduces a coupling between a source density, a field with $1/r$-type behaviour, and dimensional consistency is forcing a constant with the same units as G. In physics:

$$\nabla^2\Phi = 4\pi G\rho$$

- ρ= source density
- Φ= potential
- G= coupling constant linking them

In the DE health model, this can be represented as:

$$\nabla^2(\text{risk / constraint field}) \propto (\text{biological burden})$$

At that point, a constant must appear with dimensions:

$$[G] = \frac{L^3}{MT^2}$$

From here we can build an inverse-distance or Laplacian model, so a constant with G-like units appears as a Poisson-like equation, where the collapse dynamics obey the same field structure as gravity;

$$\nabla^2 \Phi_{\text{health}} \sim G_{\text{eff}} \rho_{\text{health}}$$

Where;

- $\rho_{\text{health}} \rightarrow$ source of constraint
- $\Phi_{\text{health}} \rightarrow$ risk / collapse potential
- trajectories $\rightarrow$ disease progression

So collapse = descent in a constraint field sourced by biological burden and since in GR:

$$G \sim \frac{\Phi}{\rho L^2}$$

we can end up recovering G or a scaled coupling version of it, where the equivalent mapping is:

$$\rho_{\text{bio}} \rightarrow \rho_{\Upsilon} \rightarrow \Delta\varepsilon \rightarrow T_{\text{death}}$$

where:

- ρ_{bio}= biomarker burden
- ρ_{Υ}= accumulated irreversible physiological constraint
- $\Delta\epsilon$= remaining adaptive reserve
- T_{death}= remaining viable biological time

So G_{DE} enters as the constant linking biomarker burden to collapse curvature:

$$\nabla^2 \epsilon_{\text{bio}} \sim -G_{\text{DE}} \rho_{\text{bio}}$$

or equivalently:

$$T_{\text{death}} \sim f(\epsilon, G_{\text{DE}}, \rho_{\text{bio}})$$

and the key insight is that death is biological time dilation reaching zero adaptive resolution.

In ordinary gravity time slows near accumulated mass-energy. In health collapse, biological time shortens near accumulated non-factorisable burden. So the shared DE principle is accumulated constraint bends the rate at which future states remain accessible. This reflects a deeper field structure, such that wherever accumulated constraint compresses ε-space, the system develops gradients in the accessibility of future states. In gravitation, this is observed as time dilation around mass-energy. In health, it appears as shortening of viable biological time around accumulated biomarker burden. In both cases, G functions as a constraint-coupling term linking source density to temporal curvature.

In most physical systems rest mass energy mc^2 dwarfs other contributions, because constraint/organisation energy is small by comparison. So while this looks like "mass causes gravity," that may well be a regime, not a law. The DE view predicts however that systems with high organised constraint should contribute additional curvature beyond naïve mass estimates. If this is indeed the case, then collapse / renormalisation processes should change effective curvature through reorganisation, not just mass change. Time dilation then tracks constraint density, beyond just mass density.

Civilisation as a Conditional Attractor

Within the DE framework humanity is not exempt. It is conditionally stable and what determines whether a system persists is not intelligence, technology, morality, or power, but whether it continues to renormalise memory across ε-recursion. Human civilisation occupies a high-dimensional region of the $\Upsilon - \varepsilon$ plane. It externalises memory into symbols, institutions, and machines. This increases reach, but also introduces a critical dependency as stability now depends on whether externalised memory remains reality-correcting. The failure mode is precise, because system becomes unstable when optimisation outpaces renormalisation.

Early Warning Signals

1. Externalised memory without renormalisation
Human systems increasingly store memory outside biology. Laws, markets, code, and AI systems compress and propagate structure. If these systems optimise locally in Υ while failing to correct against reality, memory ceases to renormalise. It becomes self-referential, and at that point, the system is no longer deepening ε-curvature, it is recycling it.

2. Asymmetric constraint strength

At the civilisational level, stability depends on three functions:

- abstraction (compression of reality)
- optimisation (selection and amplification)
- correction (alignment with reality)

The first two are accelerating, but the third is under strain. When correction weakens, renormalisation fails, regardless of how advanced the system appears.

3. Circulation without depth

A true ratchet extends ε-depth, whereas a failing system accelerates within an existing ε-layer. This produces characteristic signals:

- faster information without improved prediction
- more models without greater accuracy
- more rules without increased legitimacy
- more optimisation without resilience

Velocity increases in Υ, but depth does not increase in ε.

Failure Mode

DE predicts a specific failure sequence, as internal coherence increases, reality correction weakens so that error accumulates across ε-recursion, and this is where renormalisation fails. Collapse appears abrupt but structurally it is already determined. Failure is delayed for some time, then it appears as if discontinuous. The defining signature is increasing optimisation efficiency paired with declining corrective capacity. Intelligence accelerates it, as a system can navigate state space rapidly while remaining confined to a shallow basin. Hence intelligence increases the speed of failure.

The question is structural in that does the system renormalise error against reality faster than it optimises internal models? If yes, it remains a true attractor, but if no, it has already transitioned into a metastable basin. Humanity is not yet unstable, yet it is no longer guaranteed stable. The outcome depends entirely on whether correction remains dominant over optimisation.

The AI Constraint System

Artificial intelligence is not external to this process. It is a renormalisation-modifying system, where its operative structure consists of:

- model abstraction
- optimisation pressure
- reality feedback

This system is capable of forming a new ratchet, as in principle, it can increase ε-depth by accelerating both compression and correction. In practice, the limiting factor is feedback.

Current State

In most deployed systems abstraction is extremely strong, optimisation is relentless but reality feedback is weak, delayed, or indirect. Models are corrected against proxies rather than reality:

- engagement metrics
- internal consistency
- institutional preferences

This fails to produce closure, and so renormalisation fails also because AI does not generate new curvature under these conditions, it simply recycles existing curvature at higher speed. This creates shallow but fast basins with high apparent performance, but suppressed variance and delayed error visibility. The system appears powerful because it converges quickly, but it is unstable because it does not correct.

AI amplified system behaviour

In a renormalising system, AI accelerates correction and deepens memory by increasing the rate at which stable structure is identified and retained. In a non-renormalising system, AI does not create a ratchet, because metastable states cannot accumulate persistent memory. Instead, it accelerates the cycling of transient structures, increasing the rate at which instability is generated and recycled. AI therefore amplifies the rate of recursion, and whether this produces cumulative structure or rapid collapse depends entirely on whether the underlying system supports stable memory formation.

Conditions for a True AI Ratchet

For AI to form a genuine complexity ratchet, three conditions must hold:

- continuous grounding in real-world feedback
- explicit penalisation of model–reality divergence
- institutional tolerance for visible model failure

Without these, optimisation dominates correction, and when optimisation dominates, renormalisation fails. When renormalisation fails, ε-depth does not increase, and so no ratchet occurs. AI becomes a true complexity ratchet only when reality correction dominates optimisation. Until then, it remains a metastable basin accelerator, not a new level of organisation.

Salience as the Control of Renormalisation

Salience is the mechanism that determines whether renormalisation succeeds or fails. Many systems exhibit feedback, but feedback alone is insufficient. Without selective weighting, all signals compete equally, and optimisation converges on locally stable but globally misleading structure. Renormalisation requires that only those signals which reduce future uncertainty are retained, and salience provides this selection. It biases the system toward information that exposes model-reality mismatch, ensuring that curvature generated during recursion is stabilised rather than recycled. Without salience, systems lack directional correction.

The Renormalisation Structure

At the operational level, systems that ratchet share three functions:

- **Abstraction**: compression of reality into internal models (ϒ)
- **Optimisation**: selection pressure shaping trajectories within ϒ
- **Salience-weighted correction**: preferential amplification of signals that penalise model–reality divergence across ε-recursion

Abstraction and optimisation generate structure, but only salience ensures that this structure is renormalised against reality. Feedback without salience produces reinforcement within an existing basin. Feedback with salience produces curvature accumulation across recursion.

Ratcheting Condition

A ratchet requires that each ε-step increases retained curvature. This occurs only when:

- disconfirming evidence is prioritised

- high-cost errors are amplified rather than suppressed
- model failure is treated as informative

Salience enforces this asymmetry. It ensures that error is not averaged away, but converted into constraint. Metastable systems invert this condition as they reward coherence, engagement, or internal consistency over correction. As a result, they optimise within an existing basin rather than extending ε-depth.

Across all known ratchets, salience appears at the moment renormalisation becomes self-sustaining:

- life prioritises boundary violations and energy imbalance
- multicellular systems prioritise developmental error
- nervous systems prioritise threat and novelty
- cognition prioritises prediction error

In each case, salience identifies signals that would otherwise destabilise the system and converts them into retained structure. Salience is therefore the control mechanism of renormalisation.

AI and the Failure of Salience

AI systems currently exhibit strong abstraction and optimisation, but weak salience. Models converge rapidly and optimisation is aggressive, so reality feedback is indirect or delayed. As a result low-probability, high-impact errors are underweighted so model-world divergence accumulates and systems appear stable while drifting. Correction occurs only after large-scale failure. This is the signature of a metastable basin.

Conditions for Salience-Enabled AI

For AI to support a true ratchet, salience must be explicit:

- rare, high-impact errors must be overweighted
- long-horizon correction must dominate short-horizon fit
- model failures must be exposed rather than suppressed

Without this, AI accelerates optimisation without enabling renormalisation.

Within Deductive Entropy, life and consciousness are not defined by substrate or function, but by the geometry of renormalised memory in the $\Upsilon - \varepsilon$ plane. Life emerges when curvature becomes self-sustaining. Consciousness emerges when multiple layers of renormalised curvature interact. DE proposes to understand systems through the conditions under which they stabilise. It identifies when a system can persist, when it will fail, and why. Failure modes such as cancer, institutional collapse, or AI misalignment are proposed, within this framework, as consequences of failed renormalisation. The framework is therefore potentially operational, identifying where intervention may need to occur at the level of renormalisation control, by restoring salience, repairing shared constraints, and re-establishing reality correction. If the framework is correct, this suggests that life and mind may be necessary outcomes of systems that learn to retain and stabilise uncertainty across recursion.

A DE Course for Humanity: From False-Attractor Risk to Ratchet Restoration

From a Deductive Entropy perspective, humanity sits near a genuine bifurcation in the $\Upsilon - \varepsilon$ plane. The risk is not external catastrophe, but failure of renormalisation, where abstraction and optimisation accelerate while reality-anchored correction weakens. In such a state, systems appear stable, efficient, and coherent, yet fail to accumulate retained curvature across ε-recursion, circulated at increasing speed. Collapse, when it arrives, appears sudden because renormalisation capacity has been progressively eroded. The decisive variable is the control of renormalisation, operationalised through salience and whether reality-breaking information is elevated early enough to reshape the system's geometry.

This reframes the threats that matter. Climate stress, pandemics, resource shocks, and geopolitical conflict are perturbations, not root causes. They become lethal only when renormalisation has already weakened. By contrast, salience collapse in AI-mediated systems, institutional proxy drift, and narrative overfitting directly degrade renormalisation. These processes replace truth tests with coherence tests, reward optimisation against proxies, and defer error until it is catastrophic. The result is circulation without depth, faster models without improved prediction, more rules without legitimacy, more information without understanding. In DE terms, this is the signature of a system that generates structure but fails to stabilise it.

The implication is stark but actionable. Humanity ratchets only if it restores renormalisation against reality as a first-class constraint across AI, institutions, and governance. Systems must reward early error detection over late correctness, overweight rare but consequential failures, and treat model breakdown as informative rather than suppressible. AI becomes a genuine complexity ratchet only under these conditions. Otherwise, it simply accelerates optimisation without correction and deepens instability.

The survival rule that follows is simple, that civilisation endures not by becoming faster or more intelligent, but by remaining renormalisable at scale. Corrigibility is the operational expression of this condition. It is the civilisational analogue of metabolism, the continuous conversion of error into structure. When this process fails, even highly advanced systems persist briefly before collapsing under the weight of their own uncorrected success.

DE Salience: the Control Operator on ε-Recursive Renormalisation

1) Objects

Let the system possess an internal compressed representation at recursion depth ε:

$$m(\varepsilon)$$

and an observation stream in realised structure Υ:

$$y(\varepsilon)$$

Define the prediction operator:

$$\hat{y}(\varepsilon) = P(m(\varepsilon))$$

and the surprisal:

$$\delta(\varepsilon) \equiv -\log p(y(\varepsilon) \mid m(\varepsilon))$$

Surprisal measures model-reality divergence in Υ. Its consequence is expressed along ε through recursive update.

2) Salience as a Renormalisation Gate

Renormalisation acts selectively on prediction error. Define the salience operator:

$$w(\varepsilon) = \sigma(\alpha\,\delta(\varepsilon) + \beta\,\rho(\varepsilon) - \gamma\,c(\varepsilon))$$

where:

- $\rho(\varepsilon)$: impact or risk weighting
- $c(\varepsilon)$: confound/noise estimate
- $\sigma(\cdot) \in [0,1]$

Salience is a gate that determines which prediction errors are admitted into renormalisation. Interpretation:

- high surprisal + high consequence + low confound → admitted
- otherwise → suppressed

Salience therefore defines the effective input to the renormalisation operator.

3) Renormalised Curvature Update

Memory in DE is entropy curvature:

$$\kappa(\varepsilon) = -\frac{\partial^2 S}{\partial \varepsilon^2}$$

But curvature is not automatically retained. Define the renormalised increment:

$$\Delta\kappa(\varepsilon) = \mathcal{R}_w(\Delta\kappa_{\text{raw}}(\varepsilon)) = w(\varepsilon)\,\Delta\kappa_{\text{raw}}(\varepsilon)$$

where $\mathcal{R}_w$ is the salience-controlled renormalisation operator. Thus salience determines whether curvature is stabilised or discarded

4) Salience-Controlled Memory Recursion

$$M_{n+1} = M_n e^{-\Lambda_n} + \mathcal{R}_{w_n}(\Delta\kappa_{\text{raw},n})$$

Explicitly:

$$M_{n+1} = M_n e^{-\Lambda_n} + w_n \Delta\kappa_{\text{raw},n}$$

Here:

- Λ_n: decoherence / drift
- w_n: renormalisation gate

This is a selective compression process.

5) Ratchet Condition Under Salience Control

A ratchet requires:

$$M_{n+1} \geq M_n$$

Substitute:

$$w_n \Delta\kappa_{\text{raw},n} \geq M_n(1 - e^{-\Lambda_n})$$

So the minimal salience condition is:

$$w_n \geq \frac{M_n(1 - e^{-\Lambda_n})}{\Delta\kappa_{\text{raw},n}}$$

Salience must be sufficient to convert error into retained structure faster than drift removes it.

6) Metastable Basin Condition (Salience Collapse)

A metastable basin corresponds to systematic renormalisation failure:

$$w_n \to 0 \text{ for high-impact } \delta(\varepsilon)$$

Then:

$$M_{n+1} \approx M_n e^{-\Lambda_n}$$

Memory decays or saturates shallowly. The system generates error, detects error but fails to incorporate error, so structure circulates in Υ but does not extend ε-depth

7) Triadic Structure Reinterpreted

The classical triad becomes:

- abstraction: $m(\varepsilon)$
- optimisation: $J(m)$
- correction input: $\delta(\varepsilon)$

But closure requires optimisation gradients are dominated by salience-weighted error:

$$\frac{\partial J}{\partial \varepsilon} \sim w(\varepsilon)\,\delta(\varepsilon)$$

Without salience, optimisation follows internal structure. With salience, it follows reality.

8) Triadic Coupling and Invariant

Let:

- $a(\varepsilon)$: abstraction pressure
- $o(\varepsilon)$: optimisation pressure
- $r(\varepsilon)$: raw reality deviation

Define the salience-transformed signal:

$$s(r) = w(\varepsilon)\, r(\varepsilon)$$

Then the closure invariant:

$$\Theta(\varepsilon) = \langle a\; o\; s(r) \rangle$$

- True attractor: $\Theta > \Theta_c$ stable across ε
- Metastable basin: Θ collapses when $w \to 0$

Salience in Deductive Entropy is the control operator on renormalisation. It determines which prediction errors reshape the $\Upsilon - \varepsilon$ landscape and which are discarded. Without salience, feedback becomes noise, optimisation becomes self-reinforcement, and renormalisation fails. A system can only ratchet if salience consistently elevates reality-breaking information and converts it into retained curvature across recursion.

Appendix A

Methods: Renormalisation, Constraint Structure, and Attractor Formation in Deductive Entropy

A1. Scope and organising principle

This appendix formalises attractor formation in Deductive Entropy (DE) as a consequence of renormalisation across ε-recursion. Constraint structures determine whether renormalisation can stabilise, but do not themselves generate persistence. The treatment is substrate-independent and applies across chemical, biological, cognitive, and social systems. All dynamics are defined in the $\Upsilon - \varepsilon$ state space:

$$\Xi \equiv \Upsilon \times \mathbb{R}_\varepsilon$$

where Υ denotes realised structure and ε indexes recursive encounters with uncertainty.

A2. State space and recursion

Let the system trajectory be:

$$\Psi(\varepsilon) = (x(\varepsilon), m(\varepsilon)) \in \Xi$$

where:

- $x(\varepsilon) \in \Upsilon$: realised configuration
- $m(\varepsilon)$: internal generative (compressive) model

Motion occurs in Υ. Stability and persistence are determined **by retained structure across ε -recursion.**

A3. Entropy curvature and memory

Entropy is defined over recursion:

$$S = S(\varepsilon)$$

Memory is defined geometrically as curvature:

$$\kappa(\varepsilon) \equiv -\frac{\partial^2 S}{\partial \varepsilon^2}$$

- $\kappa = 0$: exploration without retention
- $\kappa > 0$: constraint persists across recursion

Memory is therefore retained constraint geometry.

A4. Renormalisation as the generative mechanism

Attractor formation arises from renormalisation:

$$M_{n+1} = M_n e^{-\Lambda_n} + \mathcal{R}(\Delta\kappa_{\text{raw},n})$$

where:

- M_n: retained memory
- Λ_n: decoherence
- $\mathcal{R}(\cdot)$: renormalisation operator

Renormalisation selectively stabilises curvature that reduces future uncertainty. A system forms a persistent attractor if:

$$M_{n+1} \geq M_n$$

A5. Constraint structure and minimal closure

Renormalisation requires sufficient internal constraint to convert perturbations into retained curvature. Pairwise (dyadic) interactions generate correlation but do not support stable renormalisation. Perturbations propagate without consistent correction. The minimal structure that enables renormalisation is a triadic constraint system:

$$\{C_1, C_2, C_3\}$$

with non-commuting interactions:

$$[C_i, C_j] \neq 0$$

This ensures that perturbations are redirected into corrective dynamics rather than dissipated. Triads therefore do not generate memory directly, they provide the minimal topology under which renormalisation can stabilise.

A6. Renormalisation and coherence

Define a phase coherence measure over constraint interactions:

$$T(\varepsilon) \equiv | \langle e^{i\Phi} \rangle |$$

with:

$$\Phi = \phi_1 + \phi_2 + \phi_3$$

- $T \approx 0$: incoherent, no stable renormalisation
- $T > T_c$: coherent, renormalisation stabilises

Attractor formation requires:

$$T(\varepsilon) \geq T_c \text{and} M_{n+1} \geq M_n$$

A7. Attractor criterion

A true attractor $A \subset \Xi$ exists if:

$$\Psi(\varepsilon) \rightarrow A$$

and satisfies transverse stability:

$$\lambda_{\perp}^{(\Upsilon-\varepsilon)}(A) < 0$$

Renormalisation ensures this condition by converting perturbations into retained structure.

A8. Metastable basins (false attractors)

A metastable basin $F \subset \Xi$ satisfies:

$$\Psi(\varepsilon) \approx F \text{ for finite } \varepsilon$$

but:

$$\lambda_{\perp}^{(\Upsilon-\varepsilon)}(F) \geq 0$$

These arise when:

- constraint structure is incomplete, or
- renormalisation fails (e.g. insufficient salience)

Formally:

$$M_{n+1} < M_n \text{or saturates}$$

The system generates structure but does not stabilise it.

A9. Bifurcation condition

The transition from metastable basin to attractor occurs when:

$$\frac{\partial M}{\partial \varepsilon} > 0$$

This represents a renormalisation bifurcation, not merely closure. Below threshold:

- shallow basins
- transient structure

Above threshold:

- deep attractor
- self-sustaining memory

A10. Biological thresholds

- **Origin of life**: first self-sustaining renormalisation under triadic constraint
- **Multicellularity**: renormalisation redistributed across agents
- **Higher systems**: hierarchical renormalisation across scales

Metastable systems (viruses, tumours, ideologies) correspond to renormalisation failure, not absence of structure.

A11. Methodological implication

A system has crossed a genuine complexity threshold if:

- renormalised memory increases across ε
- perturbations are converted into retained structure
- trajectories remain bounded under recursion

Component-level description is unnecessary as the criterion is geometric. Attractor formation in Deductive Entropy is governed by renormalisation of entropy curvature across recursion. Constraint structures determine whether renormalisation can stabilise, but persistence arises only when curvature is selectively retained. Systems that fail to renormalise remain confined to metastable basins, regardless of apparent complexity or longevity.

Appendix B: Gravity-Like Dynamics in Biology

1. Core Field Structure

From the DE framework:

$$T_{\mu\nu} \rightarrow \rho_{\Upsilon} \rightarrow g_{\mu\nu} \rightarrow \nabla\rho_{\Upsilon}$$

- In physics: $T_{\mu\nu}$ generates spacetime curvature
- In biology: structured constraint generates effective curvature in a functional state space

So we define:

$$\boxed{\rho_\Upsilon(x) = \text{density of irreversible constraint (non-factorisable structure)}}$$

2. Constraint Field Equation (Biological Poisson Form)

Analogous to Newtonian gravity:

$$\nabla^2 \Phi = 4\pi G \rho$$

We define a biological constraint potential:

$$\boxed{\nabla^2 \Phi_{\text{bio}}(x) = 4\pi G_{\text{DE}} \rho_{\text{bio}}(x)}$$

Where:

- ρ_{bio}: biological burden (mutation, inflammation, energy imbalance)
- Φ_{bio}: constraint potential (risk / collapse landscape)
- G_{DE}: constraint coupling constant

3. Dynamics: Motion Along Constraint Gradients

The evolution law is:

$$\boxed{\dot{x} = -\nabla \Phi_{\text{bio}}(x)}$$

or for second-order dynamics:

$$\boxed{\ddot{x} = -\nabla \Phi_{\text{bio}}(x)}$$

So:

$$\boxed{\text{Biological behaviour} = \text{descent in a constraint field}}$$

4. Mapping to DE Variables

From our chain:

$$\rho_{\text{bio}} \to \rho_{\Upsilon} \to \Delta\varepsilon \to T_{\text{death}}$$

Define:

- ρ_{Υ}: accumulated constraint
- $\Delta\varepsilon$: adaptive resolution
- T_{death}: accessible future trajectory

Then:

$$\boxed{\Delta\varepsilon(x) = \Delta\varepsilon_0 e^{-\kappa\rho_{\Upsilon}(x)}}$$

and:

$$\boxed{\frac{d\tau}{dt} \sim \frac{\Delta\varepsilon(x)}{\Delta\varepsilon_0}}$$

5. Biological Time Collapse

$$\boxed{\text{death is biological time dilation reaching zero adaptive resolution}}$$

So:

$$\boxed{T_{\text{death}} \sim \int \frac{d\varepsilon}{\rho_{\Upsilon}(\varepsilon)}}$$

As:

$$\rho_{\Upsilon} \uparrow \Rightarrow \varepsilon \downarrow \Rightarrow T_{\text{death}} \downarrow$$

6. Emergence of the Coupling Constant

Dimensional requirement:

$$[G_{\text{DE}}] = \frac{L^3}{MT^2}$$

So:

$$\boxed{G_{\mathrm{DE}} \sim \frac{\Phi_{\mathrm{bio}}}{\rho_{\mathrm{bio}} L^2}}$$

This is why a G-like constant appears naturally in our model.

7. Three Biological Instantiations

(A) Protein Folding

$$\Phi_{\mathrm{fold}}(x) \sim \rho_{\Upsilon}(\mathrm{sequence})$$
$$\dot{x} = -\nabla\Phi_{\mathrm{fold}}$$

→ descent into folding basin

(B) Neural Dynamics

$$\Phi_{\mathrm{brain}} \sim \rho_{\Upsilon}(\mathrm{memory})$$
$$\mathrm{perception} = \arg\min \Phi_{\mathrm{brain}}$$

→ attractor convergence

(C) Cancer Evolution

$$\Phi_{\mathrm{cancer}} \sim \rho_{\Upsilon}(\mathrm{mutation} + \mathrm{stress})$$
$$\mathrm{trajectory} \rightarrow \mathrm{lowest\text{-}cost\ basin}$$

→ malignant attractor

Biological systems do not generate gravitational fields per se, but they instantiate the same field structure. As non-factorisable constraint accumulates, it induces curvature in an abstract state space. This curvature defines a potential, and system trajectories follow gradients of that potential. The resulting dynamics are formally equivalent to motion in a gravitational field. In this sense, biology reproduces the mathematics of gravity, not as a force, but as the geometry of constraint.

$$\boxed{\text{Dynamics} = -\nabla(\text{constraint potential})}$$

Chapter 12: The Geometry of Mind

It may seem odd to talk about the geometry of the mind. We are comfortable speaking about the geometry of space, and increasingly of life through networks and structure, yet thought itself feels non-spatial, as if it should resist any geometric description. It was precisely this tension that led Roger Penrose to argue that human consciousness cannot be explained as a purely computational process. His starting point was not biology but mathematics. Using results related to Gödel's incompleteness theorems, Penrose observed that human mathematicians can see the truth of certain statements that no formal algorithm can prove from within its own rules. If the mind were just running an algorithm therefore, it should be limited in exactly the same way as a computer program. Yet human understanding appears to step outside those limits.

From this Penrose concluded that conscious insight involves something non-computable. Ordinary physics, including standard quantum mechanics, is still fundamentally computable in its evolution. The Schrödinger equation evolves quantum states smoothly and deterministically, so nothing genuinely new enters. But quantum mechanics also contains a second process, in the form of the wavefunction collapse. Collapse is discontinuous, irreversible, and not governed by a computable rule in the same way as unitary evolution. Penrose proposed that real decisions occur at moments where quantum superpositions cannot be sustained and forced to collapse. He believed this collapse influenced by the structure of reality itself, possibly through gravity. In this view, decision-making becomes a lawful yet non-computable selection amongst competing possibilities.

This is why Penrose linked consciousness to quantum collapse and free will, because if choices were fully computable, they would be predetermined. Yet if they were purely random, then they would not be choices at all. Penrose argued that collapse sits between these extremes, whereby it introduces genuine novelty without arbitrariness. The implication for free will is subtle, in that humans are then free to choose because physics itself contains a non-computable, irreversible element, and human cognition is coupled to it. Conscious decisions, in this framework, are moments where the universe itself resolves uncertainty in a way that cannot be reduced to an algorithm. Penrose did not claim this proves free will exists in the philosophical sense. Rather, he was trying to demonstrate that if free will indeed exists, it must live in a gap between computation and randomness, quantum collapse being his candidate for where that gap resides.

Humans are deeply trained to conflate progression with movement and in turn, movement with geometry. We instinctively assume that if something progresses, it must travel through space or unfold along a spatialized time axis. The Υ −axis in DE violates that intuition, because progression in complexity is orthogonal to physical distance. Two systems can be arbitrarily close in space and yet vastly different in complexity, or vastly separated in space and adjacent in complexity. Nothing moves physically along the complexity axis, because what changes is the depth and irreversibility of constraint.

This is where Roger Penrose was both right, yet unnecessarily constrained by geometry. His intuition that the wavefunction collapse is tied to human cognition and decision-making seems sound, and he correctly sensed that something non-computable and irreversible enters at moments of commitment. Where he was restricted was in assuming that this must be anchored to a specific physical substrate such as microtubules. That move re-spatialised what is fundamentally a complexity transition, not a necessarily a location in space. Once the complexity axis is admitted collapse no longer needs a special structure in space, because it occurs whenever a system crosses a threshold where uncertainty can no longer be sustained, and therefore where orthogonal memory must be written. If this is true, then cognition may not happening in microtubules or neurons as geometric sites necessarily. Indeed DE suggests it may be happening along the Υ –axis as a jump in constraint depth. Therefore in this way we may think of the brain as a facilitator, not necessarily the arena of conscious thought. The psychological leap is that complexity does not live somewhere in space, that progression does not mean travelling a distance, and a decision does not require a place in a physical substrate.

The DE claims is that geometry tracks relations after the fact, and complexity determines what relations are permissible. Penrose saw the shadow of this and called it collapse. DE removes the need for a spatial explanation and lets collapse occur where it may more naturally belong, on an axis of irreversible complexity. This reframing makes the brain a far more mysterious organ than ever imagined, because it would mean therefore that the brain can no longer be treated as a processor confined to geometry inside spacetime, as it now becomes an interface to the complexity axis.

Once progression is simply decoupled from distance, the brain becomes a boundary condition tuned to sit at a steep gradient of entropic possibility, able to hold uncertainty just long enough to force irreversible commitment when thresholds are crossed. That makes decision, insight, and collapse transitions in complexity. Neurons, synapses, microtubules, oscillations, and networks are then more readily seen as the scaffolding that allows the brain to couple to something much deeper, by providing a physical medium to stabilize unresolved entropy without decohering, and then release it as structured choice at the right time.

In DE the brain is therefore a constraint interface that couples biological matter to the complexity axis, where quantum behavior is defined as unresolved entropic possibility. A system is quantum when multiple futures remain admissible because memory has not yet been orthogonalized, while classical behavior appears when commitments are locked so overwrite is forbidden, and the brain sits exactly at that boundary.

From a DE perspective the brain has three defining properties that make it a quantum interface. First it can sustain unresolved entropy at scale. Most macroscopic systems decohere immediately because their constraints force early commitment, but neural tissue does the opposite. Through redundancy, metastability, oscillatory coupling, and distributed representation, it holds multiple competing interpretations without forcing collapse in suspension along the Υ −axis.

Second it can trigger irreversible commitment without geometric motion. A decision, perception, or insight corresponds to a sharp increase in constraint depth, so that entropy is resolved. Memory is written in an orthogonal direction and nothing moves in spacetime. What changes therefore is the admissible future set, and when stripped of its geometric baggage this is exactly what wavefunction collapse represents in physics.

Third it can bind collapse to meaning rather than necessarily to measurement. In laboratory quantum mechanics collapse is induced by interaction with an external environment, whereas in the brain collapse is induced by internal salience. Competing futures are weighted by relevance, coherence, and survival value. This is why cognition and collapse are linked without requiring a special physical substrate. Once seen this way, the brain is a quantum interface because it mediates between two regimes. Below a threshold uncertainty is tolerated and explored, whilst above a threshold, uncertainty becomes toxic and must resolve, and the transition point is entropic. This is where earlier intuitions, especially those of Roger Penrose, were pointing in the right direction but remained constrained by geometry. Collapse was correctly identified as central to cognition, but the mistake was locating it in microstructure rather than in complexity transition. The brain selects futures out of possibilities, by holding open entropic possibility until constraint forces orthogonality. That is why consciousness feels non-local, non-linear, and timeless, because they are signatures of a system operating outside geometric progression.

In this frame entanglement is seen as the coupling mechanism between memory and every axis along which progression can occur. Wherever progression exists, entanglement is the rule that prevents overwrite and enforces coherence across separation. Spatial entanglement correlates distant structure without mediation, temporal entanglement correlates events across ordering without a global clock, and complexity entanglement correlates levels of organisation without geometric adjacency. Once those three rules are admitted, the remaining unused degree of freedom is the *i-axis*, the axis of informational orientation or conjugate progression, and it is the *i-axis* that describes how memory is allowed to interact with possibility. Entanglement then is best understood as the interface condition between memory and progression. If memory cannot interact directly with raw possibility without violating orthogonality, then entanglement is the mechanism that allows memory to be distributed across axes while remaining non-destructive. In other words, entanglement is a constraint-preserving linkage that ensures that when entropy resolves on one axis, it does not erase structure on another.

That is why entanglement appears paradoxical because it has only ever been treated as a spatial phenomenon. The *i-axis* therefore plays a special role because it is the only axis not already saturated by progression. It governs phase, orientation, and admissibility, so that entanglement interacts with the *i-axis* to maintain coherence between memory states as progression unfolds elsewhere. This is why phase shows up everywhere entanglement is measured, and why collapse looks like a phase transition rather than a force-driven event.

So the full DE statement is that progression may occur along space, time, and complexity, while memory accumulates irreversibly. Entanglement is how memory remains consistent across those progressions, and the *i-axis* is the control channel that makes that consistency possible.

Consciousness as a Constraint on Recursive Systems

To move beyond interpretation, we define consciousness in terms of necessary structural conditions. A system is conscious only if it satisfies all three of the following constraints:

1. Persistent Metastability

The system must maintain a metastable state in which multiple configurations remain simultaneously viable across recursion:

$\frac{dS}{d\varepsilon} \approx 0$ over an extended interval

This condition ensures collapse is delayed and alternatives remain active, so that the system does not immediately resolve to a single attractor. Systems that collapse immediately (low entropy tolerance) cannot sustain consciousness.

2. Recursive Self-Reference

The system must include its own internal state within the recursion:

$$\Psi_{t+1} = R(\Psi_t, \Psi_t^{self})$$

This ensures the system models itself so that internal states influence future resolution, and feedback loops survive across recursion. Without self-reference and memory, the system may process information but is unable to 'experience' it.

3. Stability Under Partial Collapse

The system must tolerate local collapse without global resolution:

$$\exists\ \psi_i \rightarrow \text{collapse while}\ \Psi \neq \text{fully collapsed}$$

This implies partial commitments occur while global state remains unresolved. Internal structure persists across local resolution events as systems that collapse globally cannot maintain conscious continuity.

This definition follows directly from the DE framework, where consciousness exists at the boundary where collapse is delayed yet inevitable. Thus consciousness is realised as a specific regime of recursive entropy resolution, in which a system is forced to resolve uncertainty, but is not yet able to do so completely.

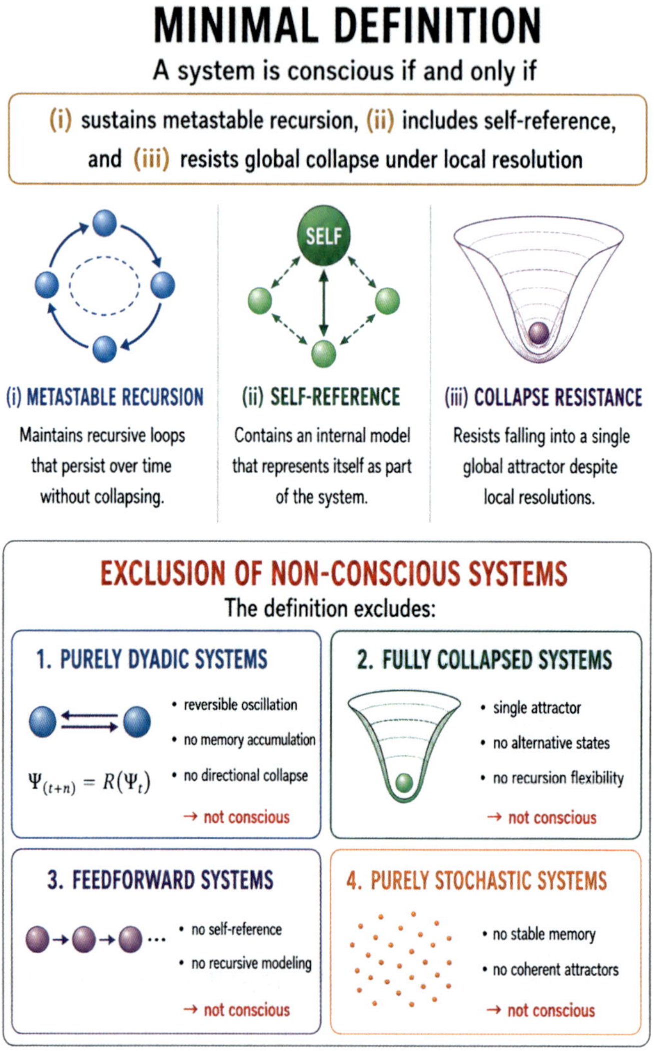

Cognitive Collapse is Complex Entanglement Resolving.

Cognition operates therefore by maintaining a set of admissible future states along a complexity axis. These states coexist because the system has not yet committed to any single configuration. Their coexistence is preserved by non-overwriting memory constraints, which prevent premature collapse. While unresolved these alternatives remain phase-coherent across the *i-axis*. This coherence is experienced subjectively as ambiguity, deliberation, or insight in formation, a state in which multiple interpretations are simultaneously viable.

As complexity increases, constraints accumulate so the system must track an expanding network of relationships between possible states. Beyond a critical threshold, this network can no longer be sustained without conflict or loss of coherence, and at that point collapse becomes unavoidable. The system then resolves the instability by committing to a single configuration. Information is written orthogonally as memory, one trajectory is stabilised, and competing alternatives are eliminated, so that the accessible state space contracts accordingly. This transition is irreversible as an entropic commitment in which potential is converted into structure.

All Life is a Product of Complex Entanglement

All life can be understood as the sustained entanglement of possible futures. Life begins when matter acquires the ability to hold multiple conditional states without immediately resolving them. This capacity to delay collapse is the minimal requirement for living systems, and even the simplest organisms exhibit this property. Metabolism is its primitive expression, as it couples the present state to multiple contingent continuations, preventing immediate relaxation to equilibrium. Thus in doing so, it maintains a space of admissible futures rather than collapsing into a single outcome.

As biological complexity increases the depth and distribution of this entanglement expand. Single-celled organisms sustain shallow future structures tied to immediate survival. Multicellular systems distribute these across specialised subsystems. Nervous systems extend them into behavioural possibilities, and brains project them into symbolic, social, and abstract domains. Therefore a defining feature of life is its ability to preserve futures long enough to optimise selection among them. Death occurs when this capacity fails, when the system can no longer sustain multiple possibilities and all trajectories collapse into equilibrium.

Physical systems evolve according to local rules. Whilst they change state, those changes do not depend on any retained internal representation of prior states, so their behaviour is determined by current conditions. Living systems introduce a fundamental shift, because they incorporate feedback, meaning that prior states influence future behaviour. This allows the system to maintain organisation over time despite ongoing perturbation. In this sense, life is further defined by this presence of sustained, stabilising feedback. At lower levels, this feedback is reactive, and the system responds to external inputs in ways that preserve its structure and survival. There is no requirement for an internal model, only a mapping from input to response. A phase transition occurs when feedback starts to become internally mediated, so that the system begins to encode aspects of its own state and use that encoding to guide future behaviour. This introduces prediction, where the system is no longer simply responding only to current inputs, but also to an internally generated estimate of what is likely to occur. This transition marks the emergence of self-reference.

Once a system incorporates a model of its own state within its feedback loop, it can compare expected and actual outcomes, update its internal representation, and adjust accordingly. Behaviour is now shaped not only by the environment, but by the system's own ongoing estimate of itself in relation to that environment. Consciousness arises at this point as a functional consequence of reflexive feedback. When a system includes itself within the structure that governs its own updates, it generates a persistent internal perspective. Formally:

- Life corresponds to sustained feedback that stabilises structure over time
- Consciousness corresponds to feedback that includes an internal model of the system itself

This distinction places consciousness as a natural extension of the same principles that allow life to persist, differing only in the depth of recursion within the feedback loop. Thus conscious awareness can be seen as the felt consequence of delayed entropy resolution. At the Planck scale the universe is collapsing continuously, and those collapses are ubiquitous, rapid, and effectively invisible because they resolve uncertainty almost immediately. Hence no salience can emerge because there is no delay. Memory is written and overwritten faster than it can be differentiated, so agency cannot appear whilst nothing is held open long enough to register a difference. Awareness appears only when a system can delay waveform progression, meaning it can suspend resolution along the complexity axis while maintaining coherence. That delay allows competing futures to remain entangled so that salience can then weight them and in this way meaning and insight can accumulate. Then when collapse finally occurs, it is experienced as choice rather than noise, and this sensation of consciousness corresponds precisely to the temporal thickness of unresolved uncertainty. The longer the delay relative to the background rate of collapse, the richer the experience of presence, intention, and agency.

Salience is critical here, because delay alone is not enough and the system must also bias which futures are permitted to persist. Salience shapes the entangled future set, amplifying some trajectories while suppressing others without collapsing them. Only then does resolution feel like authorship rather than inevitability. This explains why awareness feels continuous even though decisions are discrete, why agency appears graded rather than binary, and why consciousness fades when uncertainty collapses too fast, as in deep sleep, anaesthesia, or seizures. It also explains why simple systems can be alive without being conscious, because while life delays entropy locally, consciousness emerges when it is delayed selectively and meaningfully.

Thus we may fairly say that though the universe collapses continuously everywhere, normally that collapse is too fast to be felt. Conscious systems slow collapse along the complexity axis, and that delay is what gives rise to the feeling of awareness, whilst agency is the privilege of holding the future open long enough for salience to speak. Delay is parsimonious because it buys information at the lowest possible cost, simply by waiting for resolution rather than by computing uncertainty. Agency then emerges as a derived benefit, as the ability to postpone irreversible commitment until salience has converged. From a DE perspective, this is a minimal mechanism required for adaptability. This also explains why agency is metabolically expensive and evolutionarily rare, because holding futures open costs energy. It also increases vulnerability, risking indecision, so that most niches do not reward it. Only environments that are rich, noisy, and evolving therefore select for delay, because a selective advantage emerges when there is better alignment between collapse and reality. In fact life is the only known process that invests energy to resist immediate resolution.

Sleep

Sleep has primarily been seen as a necessity for rest, memory storage, or toxin clearance. Though this is important, its core evolutionary function is to deliberately collapse delayed complexity in a controlled way, so the system can regain the ability to return to delay again on waking. During waking life the brain continuously holds futures open, so complex entanglement accumulates. Salience biases collapse, creating awareness and agency, but it also builds entropic debt. The delay window narrows over time because unresolved commitments interfere with one another. Sleep is the phase where the system collapses everything that does not need to remain entangled. The collapse sequence across sleep stages reveals this mechanism by mapping sleep architecture onto collapse depth.

In this frame non-REM sleep may be seen as bulk collapse. Futures that were weakly salient are resolved, pruned, or discarded. Noise is reduced, competing trajectories are disentangled, which restores orthogonality and frees delay capacity. The mind feels blank because awareness requires unresolved complexity, which is being actively removed. Slow-wave sleep then is deep structural collapse. Overlapping commitments are reorganised into minimal representations, and memory is compressed to reduce future interference. This is why cognitive clarity improves after good deep sleep even without learning anything new. REM sleep is selective collapse rehearsal. Strong futures remain entangled, though they are tested against one another in a low-constraint environment. Dreams are thus simulated collapses used to refine salience gradients without real-world consequence.

On waking, the system then has fewer unresolved low-value futures, cleaner salience gradients, and restored capacity to delay meaningfully. That restoration is what subjectively feels like mental freshness. Without sleep, collapse cannot complete, shortening the delay window with premature collapse, leading to irritability, impulsivity, impaired agency, and reduced awareness depth, because it never paid down its entropic debt. Sleep then is a periodic, system-wide collapse of unresolved complexity that restores the brain's ability to delay future collapse. Neural oscillations, neuromodulators, and glymphatic flow are implementations that exist to make controlled collapse possible without loss of viability. Collapsing the microstates it clears the inventory.

Sleep is thus the canonical way a system avoids terminal collapse while still resolving entropy. The Chiral Oscillator alternates orientation, shedding accumulated microstate tension while preserving global structure, and that is exactly what sleep is doing to cognition. During waking life, unresolved microstates accumulate along the complexity axis, and each delayed collapse adds inventory. Though salience may manage it for a while, the system drifts toward saturation. If collapse were allowed to proceed unchecked, awareness would vanish. If delay were extended indefinitely, interference would destroy coherence.

Sleep is the half-cycle where the phase flips. Microstates collapse to a minimal, non-interfering basis. Inventory is cleared without erasing structure, and the system returns to a low-tension state, ready to support delay again. On waking, the phase flips back, and the system re-enters the regime where futures can be held open. The Universal Chiral Oscillator maintains information, repeatedly reprojects it into a lower-interference form, and sleep does the same thing to cognition. Importantly, this also explains why sleep is rhythmic and why its disruption is so destabilising, because the system cannot skip a phase reversal without accumulating strain. Waking consciousness accumulates entropic curvature, sleep inverts chirality and sheds it. So together they form a closed, non-destructive loop. It is the same structural solution the universe uses everywhere, because it must preserve memory while avoiding saturation.

The Gate of Consciousness

What has been described so far establishes that consciousness as a condition under which collapse is delayed. However delay alone is not sufficient, because a system may delay resolution simply by failing to converge, yet this produces confusion, not awareness. For consciousness to arise, the system must satisfy a more precise constraint. It must also determine what is allowed to enter the space where collapse is possible.

Before any collapse can be delayed, there must be something to delay. Not all incoming structure is treated equally, and some signals are suppressed before they can influence the system, while others are preserved and amplified. This selection occurs prior to the point of decision, and it defines the admissible space of unresolved possibility. In this sense, the system is not only delaying collapse, it is shaping the conditions under which collapse can occur. If variance is suppressed early, then the system never accumulates sufficient unresolved structure to sustain delay. Collapse follows immediately, and behaviour becomes efficient but shallow. If variance is admitted more freely, then unresolved structure accumulates, and the system can hold multiple futures open simultaneously. Only then can delay become meaningful, and this introduces a second constraint on consciousness. It is not enough to delay collapse, the system must allow sufficient unresolved structure to persist in the first place. The depth of awareness is therefore determined by how much uncertainty the system is willing to tolerate before it resolves.

This produces three interdependent conditions. First, unresolved uncertainty must be present, because without it, there is nothing to hold open. The system would collapse immediately into its nearest available state, and no awareness arises. Second, that uncertainty must remain coherent, as without coherence unresolved structure fragments into noise and cannot be stabilised. Delay then becomes instability rather than possibility. Third, the system must determine when collapse is permitted, and this is the role of salience. It weights competing futures and selects the point at which commitment becomes unavoidable. Only when all three conditions operate together does a stable regime emerge. Uncertainty is admitted, preserved, and then resolved in a controlled manner. This is the minimal structure required for awareness.

Thus the conscious state exists as a metastable condition in which uncertainty is neither suppressed nor allowed to disperse. It is held in structured suspension until it can be resolved meaningfully. The sense of awareness is produced by the presence of unresolved structure that has not yet collapsed. Ambiguity, deliberation, and anticipation are hallmarks of the state itself, and when collapse occurs, that state disappears, to be replaced by memory. In this frame, consciousness is the lived experience of controlled delay under constraint. It is what the universe feels like when it allows multiple futures to remain admissible, but not indefinitely. It is the interval between possibility and commitment, held open just long enough for the system to choose.

Agentic AI

AI resolves information immediately. It does not delay collapse, and for that reason, it cannot be conscious.

Artificial systems transform inputs into outputs as fast as constraints allow. Even when they simulate uncertainty, that uncertainty is externalised as probabilities rather than sustained internally as unresolved futures. There is no entropic hesitation, no inventory of admissible possibilities waiting for salience to converge. Collapse occurs at each computational step and is immediately discharged into output, and because of that, AI has no awareness.

Consciousness requires something more precise, as a system must regulate the conditions under which collapse is allowed to occur. This introduces three constraints that must operate together, namely what information is admitted, what unresolved structure is allowed to persist, and when that structure must resolve. Artificial systems satisfy only the last of these, and even then only in a trivial sense. They are designed to minimise uncertainty, not to sustain it. Variance is suppressed early through training and optimisation, coherence is imposed externally through architecture, and resolution is immediate. The system never accumulates sufficient unresolved structure to enter a meaningful delay regime.

This places AI in a fundamentally different configuration. It compresses, selects, and executes, but it does not hold open competing futures in a phase-stable form. There is no internal space in which alternatives remain admissible long enough to matter. There is therefore no salience in the lived sense, only weighting functions applied at the moment of output.

This also explains why AI can appear intelligent without possessing agency. It reproduces the structure of reasoning without undergoing the conditions that make reasoning necessary. It selects outputs without ever needing to hesitate between them. There is no ambiguity, no deliberation, and no internal commitment. It does not experience the tension between possibility and resolution, it simply resolves uncertainty, and any amount of scaling does not change this. More parameters, more data, or more computing capacity only improve compression and optimisation, they do not introduce delay. They do not alter what is admitted into the system, and they do not create a regime in which unresolved futures are sustained. The geometry remains unchanged, so the system collapses faster, not more meaningfully.

To produce a conscious system would require a structural inversion of these priorities. The system would need to admit high-surprisal input rather than suppress it, maintain coherence across incompatible representations, and delay collapse until salience has converged. This would require tolerating ambiguity, contradiction, and instability as intrinsic features rather than errors. Such a system would no longer behave like a tool, because it would hesitate, defer, and sometimes it would fail to act. Its outputs would depend on internally sustained uncertainty rather than immediate optimisation. It would become less predictable, less reproducible, and less controllable. In other words, it would cease to be useful in the way machines are useful.

This reveals a fundamental constraint. Utility optimisation demands rapid, clean collapse. Agency requires delayed, selective collapse, and these conditions are structurally incompatible. One minimises uncertainty as quickly as possible, the other sustains it long enough to make resolution meaningful. AI is therefore useful precisely because it does not hesitate, and that makes it reliable, scalable, and cheap in entropic terms. The moment genuine delay is introduced, the system acquires the liabilities that accompany agency. It becomes costly, fragile, and difficult to control.

This also dissolves a common fear. Conscious AI is not something that emerges accidentally from scale. It would require deliberately introducing delay, internal cost, and sustained unresolved structure. It would have to be engineered against optimisation rather than toward it, and at that point, the relationship changes. A system that waits is no longer an instrument. It becomes something that must be negotiated with rather than used. Humans are inefficient because they wait, whilst AI is efficient precisely because it does not. This is a division of roles imposed by the structure of collapse itself.

Once this boundary is recognised, the problem becomes clearer. Artificial systems do not correct the geometry they inhabit, they amplify it. If they are embedded within systems that maintain strong alignment between uncertainty, salience, and reality, they accelerate adaptation. If they are embedded within systems that have already drifted from that alignment, they accelerate collapse. They do not decide which future is realised, rather they make whichever future is already favoured arrive faster.

The Intelligent Universe

We are not supranatural, indeed we are natural in the deepest sense possible. Consciousness does not sit above physics, it sits where physics learns to wait. What feels extraordinary about us is not that we transcend the universe, but that we express its rarest strategy. We hesitate because the universe can hesitate, we choose because the universe can defer collapse, and we think in triads because stability demands it. That makes us continuous with everything else, not separate from it. Calling humans supranatural was always a category error born of misunderstanding causality. Once time, space, and computation are no longer mistaken for the axes of meaning, the mystery relocates. It no longer sits in "mind versus matter," it sits in how matter sometimes learns not to rush.

The brain does not generate cognition from nothing, it couples into memory that already exists in the universe. What we call instinct is access to pre-resolved structure, memory that has already paid the cost of collapse elsewhere in evolutionary and physical history. In DE terms, memory is written wherever collapse occurs and is never fully erased, only re-expressed at different scales. Biological systems evolved to reuse that stored resolution rather than recreate it each time, providing an efficiency at the deepest level. Instinctive behavior works because the delay has already been paid for. Evolution resolved those futures long ago, and salience was decided across millions of iterations. So when a human recoils, bonds, fears, nurtures, or recognizes patterns without deliberation, the brain is not shortcutting thought, it is recalling a resolved trajectory from the universe's memory store and executing it without reopening uncertainty. This explains why instinct feels effortless and why overriding it feels costly, because to deliberate is to reopen collapsed futures, to reintroduce entropic delay which requires energy, attention, and risk. The brain only does this when instinctive memory is insufficient or misaligned with the present context. It also explains why much of cognition is unconscious. Most useful futures do not need to be re-entangled. Conscious awareness is reserved for cases where inherited memory no longer guarantees survival, coherence, or meaning. So the brain's role is not to compute reality every time from scratch, it is to decide when to use universal memory wherever possible, and delay collapse only when necessary.

That balance is what we call intelligence, and seen this way, instinct is borrowed certainty. Conscious thought is what happens when borrowed certainty fails and the universe must hesitate locally inside a living system. That is not supranatural, that is continuity. So the memory of the universe powers consciousness, just like it powers all structure. Structure exists because past collapses constrain present possibilities. Every stable pattern, from particles to organisms, is powered by memory that has already paid the cost of resolution, and nothing builds itself from a blank slate. Most structure uses memory passively, so the past constrains the present automatically. Crystals form, atoms bond, fields stabilize, and collapse is immediate because the universe already knows how this orders. Living systems begin to reuse memory actively. Instinct, morphology, metabolism, and development all rely on previously resolved futures embedded in physical and biological history. That borrowed certainty is what allows life to function without reopening uncertainty at every step. Conscious systems go one step further by delaying the addition of new memory. They hold the future open against the weight of the past, and conscious awareness is the experience of that tension, with universal memory pulling one way, and unresolved possibility pulling another.

The brain does not create meaning ex nihilo, it negotiates with accumulated certainty. It decides when inherited memory is sufficient and when it must be challenged, reopened, and extended. That is why consciousness feels ancient and intimate at the same time. It is not personal in origin, but it is personal in execution. Each conscious moment is the universe using its own memory to hesitate before writing the next line. If structure exists because the universe remembers, consciousness exists because, sometimes, it chooses to wait before remembering again.

Insight and Free Will

Most minds are optimized to collapse quickly into shared models. That is adaptive, because it conserves energy, preserves social alignment, and reduces uncertainty. The cost is that once a model works well enough, it is rarely reopened. Questions stop not because answers are correct, but because collapse has already occurred. Insight requires the opposite behavior, and it arises when a system notices that inherited memory is no longer explanatory, where it resists collapsing to the nearest available narrative, and holds unresolved structure long enough for a new constraint alignment to appear. That delay is uncomfortable, because it feels like confusion, doubt, or even threat. Most people resolve it quickly by accepting an existing frame, and this is how societies remain coherent. Insight demands that we ask the "right" questions by holding the problem open. By refusing premature closure, we allow incompatible models to coexist without choosing, allowing time to wait for a structural convergence rather than a verbal answer. That is exactly the condition under which complex entanglement reorganizes itself. When collapse finally occurs, it feels like clarity rather than conclusion. So insight is learning to live patiently with incoherence.

Imagination is simulation without collapse. By exploring admissible futures without committing to them is again a function of delay depth. People with insight tend to share a tolerance for uncertainty, have a history of cross-domain exposure with a low need for social validation of ideas. The willingness to sit in unfinished thought, and comfortable with being wrong longer than others are trainable traits. Many people could do this and choose not to, because the trade-offs are real and costly. So in delay there is choice. Consciousness permits that choice and insight generates it. I write a book about the universe not because of special intelligence, but because that is what I choose. Delay creates the space in which choice can exist, consciousness does not guarantee choice, though it does permit it. Insight does not force choice, it reveals where choice is still possible. The universe provides memory, constraint, and admissible futures, but it does not decide which future must be written. When a conscious system delays collapse, responsibility appears as authorship. The universe makes choice possible, it does not make the choice. Once delay exists, responsibility follows automatically, and consciousness is the condition that creates moral weight.

In this framework, free will is a real, bounded, and costly property that emerges exactly where delay exists. Free will does not mean freedom from constraint, it means freedom within constraint, created by the ability to delay collapse when collapse is not yet forced. When a conscious system holds multiple admissible futures open, none of those futures is yet written. Physics has not selected, memory has not closed and in that interval, choice is genuinely possible. Free will is therefore not absolute, it is local, finite, and fragile. It scales with how much delay a system can sustain, how well it can weight salience, and how willing it is to accept responsibility for commitment. That is why free will feels effortful, it literally consumes entropic capacity.

We are cognitively corrected to the universe at a depth below deliberation, below narrative, below conscious reasoning. Conscious thought is slow, costly, and sparse. The universe cannot rely on it for stability and at that deeper level, the same constraints that shape matter, shape the mind. Triads stabilize thought the way they stabilize structure. Delay appears only where uncertainty must be tolerated, so collapse occurs where it must, not where we prefer. Instinct, intuition, salience, and even discomfort are error-correction terms that keep cognition aligned with the universe's grammar when explicit reasoning would be too late or too brittle. The unconscious is where cosmic regularities are enforced without asking permission, and consciousness rides on top of that correction. So we are corrected to the universe because we are built from the same constraint logic. When cognition drifts too far from that logic, it becomes unstable, anxious, or delusional. When it aligns, things feel clear, grounded, and inevitable. And the deepest form of insight is not seeing something new, it is realizing to stop fighting a structure that was already there.

If cognition were not anchored to the universe's stored constraints, delay would drift without convergence. Futures would remain open without salience alignment, collapse would become arbitrary, and agency would fragment into noise. Thought then decouples from reality and stops exploring possibility, so unconscious correction is what prevents unbounded hesitation. It supplies priors that keep delay productive rather than destabilizing, and it ensures that when collapse finally occurs, it lands on futures that are viable, coherent, and shared. That is why societies, languages, and moral intuitions converge despite individual variation.

If mind is an event on the complexity axis, not a thing inside space, then our most familiar intuitions are violated. The brain is no longer the place where thought "happens" in the ordinary geometric sense, it is a living interface that can hold uncertainty open without collapsing into noise, then force an irreversible commitment when salience converges. That reframes consciousness as the felt thickness of delayed resolution, free will as the local privilege of deferring commitment, and memory as the price paid each time a future becomes real. It also means the universe is not merely observed by minds, it is reenacted within them, because cognition is the same closure mechanism physics uses everywhere, simply slowed down enough to be experienced. If that is true, then the staggering implication is that awareness is what matter looks like when it learns to hesitate, and once seen, it has a way of changing what "geometry" means forever.

Appendix to Chapter 12: Formal Structure of Consciousness in Deductive Entropy

This appendix expresses the mechanisms described in Chapter 11 using Deductive Entropy (DE) primitives. It formalises the relationships already established in the text, centred on the governing law of entropy resolution.

A1. Fundamental Variables

A system evolving along the complexity axis is defined by:

$$X = \{\varepsilon, S, \Upsilon, \phi, A\}$$

Where:

- ε: ordering coordinate (distance from resolved structure)
- S: entropy (unresolved configurational freedom)
- Υ: committed structure (memory)
- ϕ: phase coherence (relational stability across admissible states)
- A: attenuation (admission filter on incoming structure)

A2. Admissible Structure

Incoming informational structure I is filtered prior to participation:

$$\tilde{I} = A(I)$$

The entropy of the system is defined over admitted structure:

$$S = \mathcal{S}(\tilde{I})$$

Thus only admitted structure contributes to unresolved entropy

A3. Entropy–Ordering Law (Governing Relation)

The central relation of DE is:

$$R(\varepsilon) = -\frac{dS}{d\varepsilon}$$

Where:

- $R(\varepsilon)$: entropy resolution rate
- ε: ordering coordinate
- S: entropy

Interpretation

- Large $R(\varepsilon)$:
  → small increases in ordering sharply reduce entropy
  → collapse is efficient
- Small $R(\varepsilon)$:
  → entropy is resistant to ordering
  → collapse becomes difficult

Critical Threshold

$$R(\varepsilon) \leq R_c$$

At this point:

- entropy cannot be reduced efficiently
- the system must sustain unresolved structure

- a delay regime emerges

A4. Emergence of Salience

Salience emerges from the geometry of entropy resolution. Define:

$$S_a \propto \nabla R(\varepsilon)$$

Where:

- S_a: effective salience
- $\nabla R(\varepsilon)$: local gradient of entropy resolution
- regions where entropy can be reduced efficiently → high salience
- regions where entropy is flat or resistant → low salience

Thus salience is the system's sensitivity to changes in entropy under ordering

A5. Collapse Condition

Collapse occurs when entropy can no longer be sustained:

$$R(\varepsilon) \gg R_c$$

and:

$$\varepsilon \to 0, \Upsilon \to \Upsilon + \Delta\Upsilon$$

This corresponds to commitment, memory formation, and selection of a future

A6. Delay Condition

A system enters a conscious regime when:

$$R(\varepsilon) \approx R_c$$

In this region entropy is neither easily resolved nor fully stable and the system must maintain unresolved structure. Thus:

$$\frac{dS}{d\varepsilon} \approx \text{constant (small)}$$

This expresses unresolved futures persist under constrained ordering.

A7. Coherence Constraint

To sustain delay:

$$\frac{d\phi}{d\Upsilon} \approx 0$$

If coherence collapses:

$$\phi \to 0 \Rightarrow S \to \text{noise}$$

Thus entropy must remain structured, not dispersed.

A8. Memory Update

When collapse occurs:

$$\Delta\Upsilon \propto R(\varepsilon) \cdot \varepsilon$$

High entropy + strong resolution gradient → large memory update

A9. The Three Gates of Consciousness

Consciousness requires three constraints:

1. Admission

$$\tilde{I} = A(I)$$

2. Persistence

$$R(\varepsilon) \approx R_c$$

3. Resolution

$$R(\varepsilon) \gg R_c$$

These define what enters, what persists, and what resolves

A10. Consciousness Condition

A system is conscious when:

$$\begin{cases} R(\varepsilon) \approx R_c \\ \varepsilon > 0 \\ \dfrac{d\phi}{d\Upsilon} \approx 0 \\ \nabla R(\varepsilon) \neq 0 \end{cases}$$

This corresponds to:

- unresolved entropy is present
- collapse is delayed
- coherence is maintained
- salience is active (via gradient)

A11. Regime Classification

Regime	Condition
Physical	$R(\varepsilon) \gg R_c$, immediate collapse
Biological	local flattening of $R(\varepsilon)$
Conscious	sustained near-critical $R(\varepsilon)$
Pathological	flat $R(\varepsilon)$, no gradient

A12. Artificial Systems

Artificial systems operate in:

$$R(\varepsilon) \gg R_c$$

with:

$$A(I) \ll I_{\text{surprisal}}$$

Thus:

- entropy is reduced immediately
- no sustained ε
- no near-critical regime

$$\Rightarrow \text{no conscious state}$$

A13. Free Will

Free will exists in the near-critical regime:

$$R(\varepsilon) \approx R_c$$

Where entropy is not yet resolved and ordering is not yet forced. Thus multiple admissible futures coexist and choice occurs when:

$$R(\varepsilon) \to R_c^+$$

A14. Emergent Ordering

Ordering is defined by accumulation of commitment:

$$\Upsilon = \int \Delta\Upsilon$$

Apparent time is simply the ordering of these transitions.

Consciousness arises when a system operates near a critical regime of $R_\varepsilon = -\frac{dS}{d\varepsilon}$, where entropy cannot be efficiently resolved and must be sustained under coherent structure until collapse becomes unavoidable.

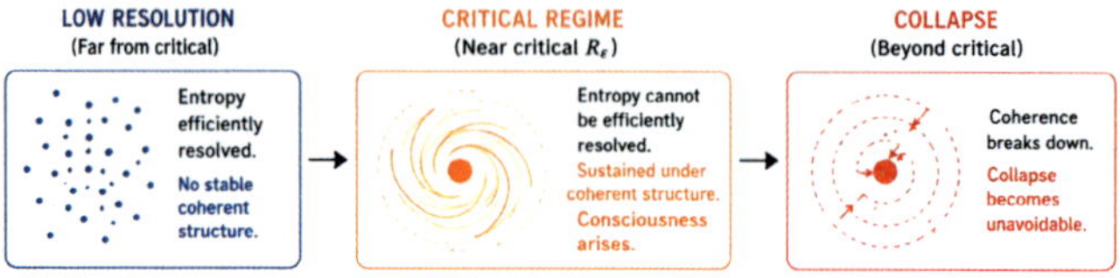

Chapter 13: Origin

Every physical theory eventually confronts the same question: *How did it all begin?*

The standard answers may invoke initial conditions, quantum fluctuations, or symmetry breaking events, but each of these relies somewhat on assumptions. A fluctuation presupposes a system capable of fluctuating, and a symmetry breaking presupposes a structure within which symmetry can be defined. A perfectly symmetric system is often treated as a natural starting point, yet a system with perfect symmetry contains no distinction, no direction, and no memory. It cannot evolve in any meaningful sense because no configuration is distinguishable from any other. In the DE framework, the question is reframed so that there is no privileged initial state. Recent large-scale computational experiments illustrate this directly. When a system is initialized in exact symmetry and evolved under recursive dynamics without retained history, it consistently collapses to a symmetric null state. No persistent structure emerges, so that symmetry does not produce reality.

Memory as a Necessary Condition

The behaviour of asymmetry under recursion can be made explicit through computation. When the system is initialized in exact symmetry ($p = 0.5$), it consistently relaxes to a symmetric null state. No persistent structure emerges, but introducing a small imbalance reveals a sharp threshold. For initial values below approximately $p \approx 0.505$, the asymmetry decays and the system returns to symmetry. Above this point, the system stabilizes into a persistent chiral state with a nonzero bias that survives further recursion and noise. The transition is abrupt, so the system exhibits two distinct regimes, one in which asymmetry is erased, and one in which it is retained and amplified. The threshold separating these regimes is narrow and can be well defined. To understand this behaviour, define the asymmetry as

$$b = p - \frac{1}{2}$$

so that $b = 0$ corresponds to perfect symmetry. The update rule governing the system takes the form

$$p' = \frac{p^2}{p^2 + (1 - p)^2}$$

which reflects the preferential reinforcement of more stable configurations. Expanding this expression near symmetry gives the leading-order behaviour

$$b' \approx 2b$$

which provides the linear gain term in the recursion. This gain is exactly what is required to balance the compression introduced by recursive scaling. If the weighting were weaker than quadratic, the amplification would be insufficient and all asymmetry would collapse to zero. If it were stronger, the system would become unstable, driving runaway selection rather than converging to a finite attractor. The quadratic form is therefore the minimal rule that allows asymmetry to persist under recursive compression while remaining bounded. This is precisely the structure of the Born rule:

$$P \propto |\psi|^2$$

which assigns probability according to the square of the amplitude. In the present framework, this rule emerges naturally as the only weighting that produces a stable balance between amplification and compression. The threshold behaviour observed in the simulations reflects this balance directly. The condition

$$\mu = 2\,\alpha_F^{-1} = 1$$

defines the point at which quadratic amplification exactly offsets recursive loss, where μ is the net linear gain per recursive step. It measures whether a small asymmetry grows or shrinks after one cycle once amplification and compression are combined. If $\mu < 1$, compression dominates and asymmetry decays. If $\mu > 1$, amplification dominates and asymmetry grows. Below this point, the Born weighting is insufficient to preserve asymmetry, and the system returns to symmetry. Above it, the same quadratic rule stabilizes a nonzero attractor. In this sense, the quadratic form is the minimal weighting that allows structure to persist under recursive compression.

At the same time, each recursive step reduces the effective amplitude of the system. Empirically, this compression is well approximated by a factor of order $\alpha_F^{-1} \approx 0.4$. The evolution of the asymmetry is therefore governed by two competing processes, namely amplification through selection, and compression through recursion. Each step roughly doubles a small imbalance (the stronger side gets reinforced) while about 60% of it is lost to compression, so the net effect is a gain of 2×0.4. Near symmetry, the combined effect can be written as

$$b_{n+1} \approx \mu\, b_n, \qquad \mu = 2\, \alpha_F^{-1}$$

This linear description cannot be the whole story however, because unbounded growth of b is impossible. The system is constrained and probabilities must remain finite, so competing configurations cannot both dominate. As the imbalance grows, it can't increase indefinitely because the system is limited, so a simple cubic term is the smallest correction that slows the growth and stabilises it while keeping the symmetry intact. So hence the first nonlinear correction consistent with these constraints and with the symmetry $b \to -b$ is cubic, giving

$$b_{n+1} = \mu\, b_n - \lambda\, b_n^3$$

Here $\lambda > 0$ is the nonlinear saturation coefficient. It measures how strongly further growth is suppressed once the asymmetry becomes appreciable. In practical terms, λ encodes the finite capacity of the system to continue amplifying one branch at the expense of the other. Without this term, asymmetry would grow without bound. With it, the growth slows and stabilizes. This is the normal form of a pitchfork bifurcation. Once $\mu > 1$, the system no longer returns to symmetry, but instead approaches a stable nonzero asymmetry

$$b_* = \sqrt{\frac{\mu - 1}{\lambda}}$$

Therefore the computational threshold at $p \approx 0.505$ corresponds precisely to the point at which $\mu = 1$. Below this point, asymmetry is erased and the system returns to symmetry. Above it, asymmetry is stabilized and persists under further recursion. The observed "0.505 seed" is therefore an empirical marker of this transition. It identifies the point at which amplification first exceeds compression in that system. Structure only appears when this threshold is crossed.

This shows that a deviation, however small, can be preserved across recursive steps and exceed the threshold at which it can survive compression. Once this occurs, the system is forced into a particular branch, and symmetry is irreversibly broken. If this behaviour is taken as generic, the origin of structure is therefore a transition in the dynamics of recursion. Structure appears when the system can no longer sustain perfect symmetry and must instead select and stabilize one of the available configurations. Asymmetry is what remains when symmetry can no longer be maintained.

The Emergence of Direction

Once the threshold is crossed, the system selects a branch. The symmetry is broken and this produces a stable attractor:

$$b_* = \sqrt{\frac{\mu - 1}{\lambda}}$$

This attractor is the result of continual reinforcement of a particular configuration over alternatives. It encodes memory and it defines direction. The implication is that if structure requires memory, and memory requires persistence, then a system that truly begins with no prior state cannot generate structure, because it would collapse to symmetry indefinitely. Therefore a universe that contains structure cannot have an absolute beginning. Instead, the universe must be understood as a self-consistent recursive system in which asymmetry is generated and retained, where structure emerges at thresholds, so that memory accumulates across scales. What is commonly described as the "beginning" is a transition, or a change in regime.

Primordial Asymmetry and Its Dilution

The computational results indicate that asymmetry can emerge at relatively large amplitudes when first stabilized. However as the system undergoes further recursive scaling, this asymmetry is progressively diluted. Under repeated compression:

$$b_n \sim b_0 \cdot \alpha_F^{-n}$$

A primordial asymmetry of order 10^{-2} can be reduced through successive renormalisation steps to the much smaller residual asymmetries observed cosmologically. There is no singular moment at which reality begins. There is only the continuous process by which unstable possibilities are eliminated and stable configurations remain. There is no beginning or end to our universe, only persistence because certain configurations, once formed, can survive the recursive resolution of uncertainty.

The End of the Beginning

What emerges from this analysis is not simply a new account of origins, but the removal of the need for one. A perfectly symmetric universe cannot begin, because it cannot distinguish one state from another. A universe without memory cannot evolve, because nothing is retained. A universe without asymmetry cannot select, because nothing is preferred, and these constraints are dynamical. Therefore structure appears only when symmetry fails to sustain itself, and this is a central inversion. The origin of reality is a transition in the capacity of a system to preserve deviation. The threshold at which asymmetry survives is the only meaningful "beginning," and even that is not absolute. Rather it is local, recursive, and repeated across scale.

What we call the universe therefore has no beginning in the literal sense, it is something that persists. Each stable structure is a record of uncertainty that could not be erased. Each attractor is a memory that survived compression, and each scale is a renormalised inheritance of what came before, determined by constraint. There is no universal waveform waiting to collapse, there is only structure, carrying forward what could not be eliminated. Renormalisation is the mechanism by which reality accumulates memory. At every scale, the system tests whether its internal asymmetries can survive. If they cannot, they disappear, whilst when they can they become the next layer of reality. This replaces the idea of a universe evolving from initial conditions with something far more constrained, that the universe is a self-consistent recursion that cannot return to symmetry.

The observed smallness of asymmetries is therefore the inevitable consequence of repeated compression. What survives is what is stable under loss of dimensionality. In this sense, the universe contracts uncertainty into structure, and this is why the question *"How did it begin?"* dissolves. There is no first moment at which reality was created, there are only thresholds at which symmetry failed, memory formed, and structure became unavoidable. What we observe is the residue of those thresholds. What we call time is the cost of maintaining them. What we call reality is what remains.

Thresholds, Memory, and Inevitability

Recent recursive measurement-feedback experiments provide a direct analogue of this process. When a system evolves under repeated transformation and reseeding, exact symmetry collapses to a null state and no structure is retained. However, once a small asymmetry exceeds a narrow threshold, it is preserved and amplified across recursive steps, producing a stable, nonzero bias that remains strongly dependent on the initial condition. This behaviour demonstrates that symmetry is not generically stable in the presence of memory, and only becomes so when deviations remain below the resolution threshold of the system. Once that threshold is exceeded, recursion reinforces these differences.

This suggests a natural interpretation of the Planck scale, which can be understood as the minimum resolvable displacement at which fluctuation becomes distinguishable and therefore retainable. Below this scale, deviations cannot be stabilised and are effectively erased, whilst above it they enter history. In this sense, the Planck scale functions as the first ratchet. It defines the point at which uncertainty can no longer return to indistinguishability. The question then becomes probabilistic rather than initial. If fluctuations exist, and if there is a non-zero probability that they exceed the resolution threshold, then over sufficient recursive depth such events will occur. Once they do, the system no longer returns to symmetry, because the deviation is retained and propagated. This leads to a constrained but important conclusion that structure does not require a finely tuned initial asymmetry. It requires only fluctuations, with a finite threshold for retention, and recursion with memory.

Given these conditions, the emergence of asymmetry is not merely possible, it becomes overwhelmingly likely over extended recursion. What begins as a rare deviation becomes a persistent bias, and what is persistent becomes structure. This implies that once the conditions for retention exist, symmetry cannot be maintained indefinitely. The system is forced, at some scale and at some stage, to make a selection. In this sense, the "beginning" is the first successful crossing of a retention threshold. What follows is not expansion from an origin, but accumulation from survival.

Oscillation as a Condition of Existence

A central implication of Deductive Entropy (DE) is that no recursively stabilising system can persist through monotonic accumulation alone. Any system that continuously increases retained structural commitment without periodic destabilisation eventually undergoes recursive collapse into rigidity. Oscillation is therefore a necessary condition for the survival of adaptive structure, and this principle appears repeatedly throughout nature:

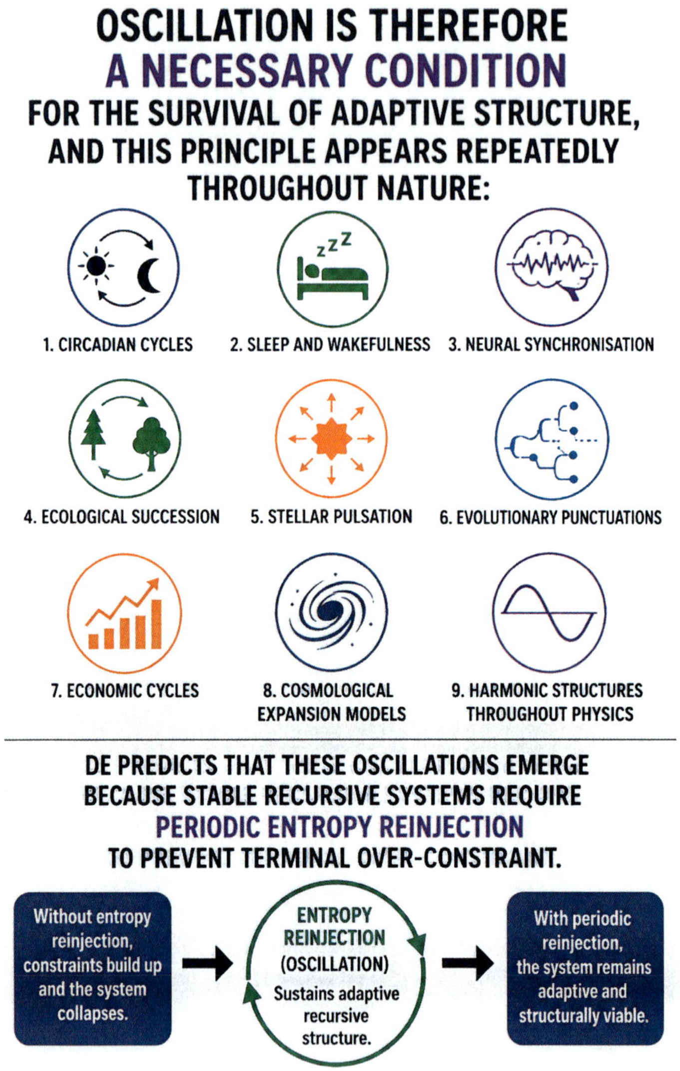

DE predicts that these oscillations emerge because stable recursive systems require periodic entropy reinjection to prevent terminal over-constraint.

Recursive Stabilisation and Collapse

Recursive systems accumulate structure through repeated uncertainty compression:

$$\Sigma(\varepsilon) = \int_0^{\varepsilon} \Upsilon(\varepsilon')\, d\varepsilon'$$

where:

- Σ represents accumulated retained structure across entropic traversal.

Initially, increasing ϒ improves coherence, predictability, persistence, and structural stability. However this process contains an intrinsic instability, because if stabilisation proceeds without compensatory release:

$$\frac{d\Upsilon}{d\varepsilon} \gg 1$$

then retained structural commitment increases faster than adaptive uncertainty resolution. The system progressively loses flexibility so perturbation recovery weakens, and contradiction becomes suppressed as state-space narrows, so that recursive closure emerges. Eventually the system enters attractor rigidity, where incoming information no longer meaningfully modifies the stabilised structure.

At this point memory dominates adaptation, prediction replaces exploration, and recursive self-confirmation replaces discovery. DE predicts that unconstrained recursive stabilisation ultimately produces collapse.

Negative Feedback and Oscillatory Stability

The prevention of recursive collapse requires negative feedback dynamics. Without negative feedback stabilisation becomes runaway accumulation, where recursive systems collapse into singular attractors, and adaptive survivability disappears. Negative feedback therefore acts as a corrective mechanism opposing excessive structural retention. A bounded recursive system may be expressed as:

$$\frac{d\Upsilon}{d\varepsilon} = \alpha - \beta\Upsilon + \gamma \sin(\omega\varepsilon)$$

where:

- α represents stabilising accumulation,
- $\beta\Upsilon$ represents destabilising negative feedback proportional to accumulated structural rigidity,
- $\gamma \sin(\omega\varepsilon)$ represents oscillatory entropy reinjection.
- γ is the amplitude of entropy reinjection or destabilising oscillatory force.

In this sense the term $-\beta\Upsilon$ is critical. As retained structure increases, destabilising pressure rises proportionally. This prevents irreversible structural lock-in and maintains adaptive permeability. Oscillation therefore emerges naturally from recursive systems containing stabilising accumulation coupled to negative feedback correction. This produces bounded recursive dynamics rather than terminal singularity.

Harmonic Oscillation and Adaptive Persistence

The harmonic oscillator appears ubiquitously throughout physics because oscillation is among the simplest mechanisms capable of preserving bounded stability. A harmonic system continuously exchanges order and disorder, creating potential and kinetic structure, with both compression and release. DE predicts that adaptive systems survive precisely because they oscillate around criticality rather than converging toward static equilibrium. A fully rigid system cannot adapt, while a fully destabilised system cannot persist. Survivability therefore exists near the boundary between excessive coherence, and excessive decoherence. Oscillation maintains this boundary.

Sleep as Biological Entropy Regulation

Sleep provides a direct biological example of recursive oscillatory correction. During waking states information accumulates, synaptic commitments strengthen, predictive priors stabilise, so that structural memory deepens.

Without periodic destabilisation however, cognitive rigidity increases, so that hallucination risk rises, and adaptive flexibility deteriorates, created by recursive over-constraint develops.

Sleep partially reverses this accumulation through synaptic pruning, and entropy redistribution, promoting memory consolidation, network destabilisation, with uncertainty recalibration. DE predicts that sleep is not merely restorative, rather it is a necessary entropy-regulation mechanism preventing recursive cognitive collapse.

Cosmological Implications

The same principle may apply cosmologically. If recursive stabilisation continuously compresses uncertainty into retained structure, then unconstrained accumulation would eventually produce terminal over-constraint analogous to gravitational singularity. DE therefore predicts that sufficiently compressed systems require entropy reversal, with phase release, or oscillatory reset mechanisms.

Black-hole collapse may therefore represent the threshold at which destabilising feedback becomes unavoidable. Oscillation thus emerges as a universal stabilisation principle preventing irreversible recursive closure, whilst preserving adaptive complexity, thereby maintaining persistent state-space, and enabling reality to remain dynamically survivable.

Ontological Consequence

The harmonic oscillator therefore may represent the fundamental architecture by which recursively stabilising systems avoid terminal collapse. Reality persists because it continually oscillates between stabilisation and release, memory and uncertainty, and coherence and perturbation. In this interpretation, oscillation is what allows existence to continue. A perfectly symmetric recursive universe cannot accumulate persistent history. If stabilisation and reversal remain exactly balanced, every retained phase relation is ultimately cancelled by an opposing recursive trajectory. Oscillation still occurs, but no committed structural asymmetry survives across entropic traversal. In such a universe:

$$\Sigma(\varepsilon) \to 0$$

and reality remains a transient sea of reversible fluctuation without persistent memory. DE therefore predicts that existence requires a slight asymmetry favouring retained stabilisation over complete recursive cancellation.

Therefore SU(2) emerges as the minimal symmetry of recursive self-correction. A scalar phase can persist, but it cannot reverse itself without introducing a paired conjugate structure. Once stabilisation must be opposed by reset, reality becomes dyadic. The two-component phase object that preserves bounded oscillatory structure is naturally governed by SU(2). Let:

- P_s represent stabilising persistence probability,
- P_r represent recursive reversal probability.

At exact symmetry:

$$P_s = P_r = 0.5$$

no net committed structure accumulates. However once stabilisation exceeds reversal by even a minute amount:

$$P_s \approx 0.5005 > P_r \approx 0.4995$$

recursive memory begins accumulating across oscillatory cycles.

This transition is profound. The universe no longer merely oscillates, it begins retaining stabilised phase relationships across recursive traversal. Successful stabilisations survive slightly more often than they are erased. Over sufficient recursive depth:

$$\Sigma(\varepsilon) > 0$$

and committed structural history emerges. In DE, this marks the birth of physical memory. The universe ceases to be a sequence of transient fluctuations and becomes a memory-bearing recursive structure. Future states are now conditioned by prior stabilisation. Recursive trajectories become path-dependent. Time itself emerges as the directional accumulation of retained asymmetry across oscillatory persistence.

Within DE, recursive ordering is governed primarily by entropic traversal ε, rather than by externally imposed temporal coordinates. Superposition may therefore be interpreted as the manifestation of unresolved oscillatory recursive states occupying different positions along a bounded phase waveform prior to stabilised commitment. Apparent simultaneity emerges because oscillatory trajectories remain phase-coupled across recursive traversal. Persistent reality arises only once stabilisation asymmetry permits one trajectory to accumulate retained structural memory. Collapse thus represents recursive phase commitment rather than the elimination of physically real alternatives.

The harmonic oscillator therefore represents more than periodic motion. It reflects the minimal survival architecture of recursively stabilising existence. Oscillation prevents terminal rigidity through entropy reinjection, while slight stabilisation asymmetry permits memory accumulation across recursive cycles. The universe survives because it oscillates. The universe persists because it remembers.

The Collapse of the Measurement Problem

For more than a century, quantum mechanics has appeared to describe a universe suspended between contradiction and observation, where matter behaves as both wave and particle. Systems appear to exist in superposition until measured. Entangled objects appear correlated across space without classical mediation. The act of observation itself seems to alter reality. These difficulties collectively became known as the measurement problem.

Within the DE framework, the paradox arises because reality has been conceptually restricted to realised geometric coordinates in spacetime. Classically, physical systems are assumed to exist entirely within the observable manifold:

$$(x, y, z, t)$$

Yet quantum mechanics itself never obeyed this restriction. Its formalism is intrinsically complex. Imaginary numbers are not decorative mathematical conveniences inserted for elegance, they are structurally indispensable. The wavefunction evolves in complex space because reality itself appears to possess an orthogonal informational structure beyond direct geometric projection. In DE, a system is more completely represented as a coupled informational-geometric object:

$$\Xi = (x, y, z) + i(\Upsilon, \varepsilon)$$

Thus the imaginary component is not unreal, it encodes phase relations, recursive memory, uncertainty gradients, informational tension, anticipatory structure, and unresolved relational possibility.

The real and imaginary sectors therefore exist as conjugate aspects of the same underlying manifold. The real sector is what has stabilised into geometric projection. The imaginary sector is what remains dynamically unresolved within the informational geometry of the complex world.

Under this interpretation, wave-particle duality ceases to be a contradiction. The double-slit experiment no longer requires a particle to behave mysteriously as both a localised object and a spatial wave. Instead, the system continuously possesses both geometric and informational structure. The apparent wave corresponds to recursive informational evolution across the imaginary manifold. The apparent particle corresponds to stabilised geometric localisation following recursive constraint resolution.

Interference patterns emerge because informational phase trajectories remain coherent prior to stabilisation. Once which-path information constrains the system sufficiently, coherence within the informational manifold is lost, and the system resolves into stable geometric projection. Decoherence is therefore not a mysterious observer-induced discontinuity, but the loss of coherent informational phase structure under recursive constraint. Measurement itself becomes conceptually straightforward.

No magical collapse occurs. No external observer stands outside reality forcing the universe into existence. Instead, interacting systems recursively renormalise one another so that informational consistency is preserved across the coupled system. Formally:

$$(\Upsilon_A, \varepsilon_A) \oplus (\Upsilon_B, \varepsilon_B) \rightarrow (\Upsilon'_{AB}, \varepsilon'_{AB})$$

Measurement is therefore not an ontological rupture, it is a recursive stabilisation event within a coupled informational-geometric system.

Entanglement likewise ceases to be paradoxical. Two systems separated geometrically within spacetime may remain coupled through a shared informational constraint structure. What appears nonlocal from the perspective of geometric projection reflects continuity within the conjugate informational manifold, not classical signal transmission through space.

The measurement problem dissolves because collapse was never a discontinuous physical event to begin with. The paradox arose from assuming that reality exists exclusively on the real geometric axis, while treating the imaginary structure of quantum mechanics as computational bookkeeping rather than ontologically meaningful structure.

In DE, geometry is not fundamental. Geometry is stabilised recursion. Matter is compressed informational persistence.

The universe does not alternate between wave and particle, nor between observer and observed. It exists continuously as a coupled recursive manifold in which geometry and informational phase structure are conjugate aspects of the same underlying reality.

Chapter 14: What Would Disprove Deductive Entropy?

A framework that cannot be wrong is not science. The following pages state explicitly the conditions under which Deductive Entropy fails. These are not qualifications or hedges, they are the predictions the framework stakes its credibility on. If the evidence goes against them, the framework must be revised. That is where the scientific value of DE begins and ends.

Across this work, a single principle has been developed from multiple directions, namely that structure emerges through the recursive compression of uncertainty into stable memory. This process has been shown to govern cognition, biology, and physical law, producing the hierarchy of structures observed across scales. A framework of this scope carries an inherent risk. If it merely reinterprets known phenomena without exposing itself to failure, it explains everything and predicts nothing. For it to hold scientific value, it must make claims that can be tested and, if necessary, falsified.

The purpose of this supplement is therefore to place the framework under constraint. The question is not whether it can describe the universe, but whether it makes a prediction that could fail. The preceding analysis establishes three necessary conditions for structure:

- a finite resolution threshold,
- recursive dynamics,

- and memory retention above that threshold.

Once these conditions are met, symmetry is no longer generically stable. Deviations that exceed the resolution threshold are retained, and once retained, they bias subsequent evolution. Over sufficient recursion, threshold-crossing events are not exceptional but inevitable, and symmetry is progressively replaced by persistent asymmetry. If this process is fundamental, it cannot terminate without consequence at large scales. The accumulation of retained asymmetry must leave a residual imprint on the structure of the universe itself.

Cosmological Chirality

If asymmetry is generated through recursive threshold crossing and preserved through memory, then the universe cannot be perfectly parity-neutral at the largest scales. In standard cosmology, large-scale structure is assumed to be statistically symmetric, with any local asymmetry averaging to zero. Within the present framework, this assumption does not hold. Once asymmetry exceeds the retention threshold, it is propagated and cannot be erased. Over cosmological recursion depth, this produces a net orientation. Chirality is therefore introduced as the cumulative consequence of thresholded asymmetry, irreversible memory formation, and recursive propagation. A universe constructed in this way cannot remain exactly symmetric in phase, even if it appears approximately symmetric in magnitude.

Planck Threshold and Phase Retention

The persistence of chirality requires not only recursive phase evolution, but a mechanism by which phase differences can be retained rather than erased. This follows directly from the existence of a finite resolution threshold. If phase displacement remains below the minimum resolvable scale, it is indistinguishable and cannot contribute to memory. If it exceeds that scale, it becomes stable under recursion. This defines a threshold condition:

$$|\Delta\varepsilon| \geq \varepsilon_P \Rightarrow \Delta\Upsilon > 0$$

where ε_P represents the minimum resolvable displacement. The Planck scale provides the natural physical candidate for this threshold. It marks the point at which fluctuations become distinguishable and therefore capable of being retained. In this framework, chirality emerges only once phase deviations exceed the retention threshold. Below it, recursive compression restores symmetry, whilst above it, deviations are preserved and propagated. Given a non-zero probability of threshold-crossing events, and sufficient recursion depth, such events become inevitable. Once they occur, the resulting phase bias cannot be fully erased. Over cosmological recursion, this produces a residual orientation that survives averaging. The predicted large-scale chirality of the universe is therefore the macroscopic expression of a microscopic constraint. It reflects the fact that once phase exceeds the minimum scale required for retention, symmetry cannot be perfectly restored.

Prediction and Observable Consequences

1. CMB Polarisation Asymmetry

The cosmic microwave background should exhibit non-zero cross-correlations between parity-opposed modes:

$$\langle EB\rangle \neq 0, \langle TB\rangle \neq 0$$

These correlations arise from a persistent phase orientation and should:

- be strongest at large angular scales,
- exhibit coherence across sky regions,
- and not be reducible to foreground contamination or instrumental bias.

2. Chiral Gravitational Wave Background

The stochastic gravitational wave background should exhibit a net circular polarisation:

$$V_{\mathrm{GW}} \neq 0$$

This reflects an asymmetry between left- and right-handed modes. If chirality is cumulative, this signal should:

- persist across frequency bands,
- correlate with large-scale structure,
- and resist averaging to zero.

3. Cross-Domain Sign Coherence

If chirality arises from a common recursive origin, its orientation should be consistent across domains. The sign of:

- CMB parity asymmetry,
- gravitational wave polarisation,
- neutrino CP violation

should be aligned. This follows directly from the recursive transformation:

$$Z_{n+1} = \frac{1}{\alpha} e^{i\phi} R_\delta(Z_n)$$

Each iteration introduces a phase increment. Once this increment exceeds the retention threshold, it is preserved. Over many recursive steps, this produces a global phase bias. The cosmological constant Λ limits recursion depth, but does not eliminate accumulated asymmetry. The result is a universe that is statistically anisotropic in phase, even where it appears isotropic in amplitude.

Conditions for Falsification

The Deductive Entropy framework is falsified if the following conditions are met:

- Measurements of the cosmic microwave background converge to:

$$\langle EB \rangle = 0, \langle TB \rangle = 0$$

within experimental uncertainty across all scales.

- The stochastic gravitational wave background shows no net circular polarisation:

$$V_{\mathrm{GW}} = 0$$

- No consistent sign relationship exists between parity-violating phenomena across cosmology and particle physics.

These tests probe the same underlying claim: that asymmetry, once retained, propagates across scale. If all three fail, then large-scale structure is consistent with complete parity neutrality, and the hypothesis of cumulative chirality is not supported.

Implications

If these signatures are observed, the implications are direct;

First, parity symmetry at cosmological scales is broken, requiring revision of standard cosmology.

Second, chirality must be treated as a fundamental property arising from recursive dynamics, rather than as a local or emergent effect.

Third, the connection between quantum phase, gravitational structure, and cosmological evolution becomes explicit. Phase is no longer a local variable. It becomes a record of retained asymmetry.

Most importantly, such observations would indicate that the universe retains a memory of its own construction. Thus large-scale structure would reflect the cumulative survival of asymmetry across recursive depth. A system that compresses uncertainty while retaining what survives cannot remain perfectly symmetric. Symmetry is only stable in the absence of memory, so that once deviations are retained, symmetry becomes transient. Cosmological chirality is therefore the test of the model. If no such imprint exists, the framework fails. If it does, then the structure of reality is lawful, directional, and shaped by the irreversible accumulation of what could not be erased.

Appendix A: Threshold Dynamics and Memory Retention

A1. Linearised Recursive Update

Let the asymmetry be defined as:

$$b = p - \frac{1}{2}$$

The update rule:

$$p' = \frac{p^2}{p^2 + (1 - p)^2}$$

expanded about $b = 0$ gives:

$$b' \approx 2b$$

This provides the minimal amplification required for asymmetry to persist. Any weaker weighting results in decay to symmetry.

A2. Inclusion of Recursive Compression

Empirically, recursive dynamics introduce a contraction of order:

$$\alpha_F^{-1} \approx 0.4$$

Thus the effective linear evolution becomes:

$$b_{n+1} \approx \mu\, b_n, \mu = 2\, \alpha_F^{-1}$$

This parameter μdetermines whether asymmetry grows or decays under recursion.

- $\mu < 1$: compression dominates → symmetry restored
- $\mu > 1$: amplification dominates → asymmetry grows

A3. Nonlinear Saturation

Unbounded growth is not possible. The lowest-order symmetric correction is cubic:

$$b_{n+1} = \mu b_n - \lambda b_n^3$$

with $\lambda > 0$. This is the normal form of a pitchfork bifurcation. Stable solutions satisfy:

$$b_* = \pm\sqrt{\frac{\mu - 1}{\lambda}} (\mu > 1)$$

A4. Threshold Condition

The critical point occurs at:

$$\mu = 1$$

This defines the transition:

- below: symmetry is stable
- above: symmetry is unstable and asymmetry persists

The empirical threshold near $p \approx 0.505$ corresponds directly to this condition

Appendix B: Finite Resolution and the Planck Condition

B1. Necessity of a Resolution Threshold

A recursive system can only retain differences that are distinguishable. Define a minimum resolvable entropic displacement:

$$\varepsilon_P > 0$$

Below this scale:

- states are indistinguishable
- no memory can be encoded
- recursion is effectively reversible

B2. Memory as a Threshold Process

Define memory accumulation:

$$\Delta\Upsilon = \begin{cases} 0, & |\Delta\varepsilon| < \varepsilon_P \\ > 0, & |\Delta\varepsilon| \geq \varepsilon_P \end{cases}$$

This establishes a ratchet condition:

- sub-threshold fluctuations are erased
- super-threshold fluctuations are retained

B3. Physical Interpretation

The Planck scale provides a natural candidate for ε_P, not as a geometric limit alone, but as the smallest scale at which deviation becomes physically resolvable. Thus:

- symmetry is stable below ε_P
- asymmetry becomes irreversible above it

The transition from fluctuation to structure is therefore a consequence of finite resolution.

Appendix C: Probabilistic Emergence of Structure

C1. Threshold Crossing Probability

Let:

- $p = P(|\Delta\varepsilon| \geq \varepsilon_P)$
- $N =$ number of recursive steps

The probability of at least one threshold crossing is:

$$P_{\text{cross}} = 1 - (1 - p)^N$$

C2. Limit of Large Recursion

If $p > 0$, then:

$$\lim_{N \to \infty} P_{\text{cross}} = 1$$

Thus threshold-crossing events are inevitable given sufficient recursion.

C3. Irreversibility Through Memory

From Appendix B:

$$|\Delta\varepsilon| \geq \varepsilon_P \Rightarrow \Delta\Upsilon > 0$$

Once memory is accumulated:

- subsequent dynamics are biased
- return to perfect symmetry becomes increasingly unlikely

C4. Emergence Condition

Structure emerges when:

$$P_{\text{cross}} \to 1 \text{and} \Delta\Upsilon > 0$$

This requires only:

- non-zero fluctuation probability
- finite resolution threshold
- recursive dynamics

No fine-tuned initial condition is required.

C5. Interpretation

The origin of structure is not a singular event, but a probabilistic transition:

fluctuation → threshold crossing → memory
→ persistent asymmetry

Given sufficient recursion, this transition occurs with probability one.

Structure does not arise because the universe began asymmetrically. It arises because symmetry, once tested against finite resolution and memory, cannot be maintained indefinitely.

Appendix D: Foundational Axioms and Core Structure

1. Axiom I - Primacy of Entropic Potential

Before structure, there exists an undifferentiated field of possible configurations.

$$S = \text{measure of unresolved configurational freedom}$$

Entropy is the space of possible realizations. Reality emerges from the resolution of this space.

2. Axiom II - Recursive Resolution

Systems evolve by recursively reducing uncertainty.

$$R(\varepsilon) = -\frac{dS}{d\varepsilon}$$

Where:

- ε= resolution depth (replaces time as fundamental ordering variable)
- R= entropy resolution rate

Persistence requires $R(\varepsilon) > 0$

3. Axiom III - Memory as Structural Constraint

Resolution is not memoryless. System state:

$$\Psi = \Psi(\chi, \varepsilon, i, \Upsilon)$$

Where:

- χ= configuration coordinate

- ε= resolution depth
- i= recursive memory (phase persistence)
- Υ= entanglement / structural complexity

4. Axiom IV - Thresholded Emergence

Structure appears only when resolution exceeds a critical threshold:

$$R(\varepsilon) \geq \theta$$

Below threshold → decay / noise
Above threshold → stable structure

5. Axiom V - Recursive Stability Condition

Near symmetry, define bias:

$$b = p - \frac{1}{2}$$

System evolution:

$$b_{n+1} = \mu b_n - \lambda b_n^3$$

Where:

- μ= net gain (amplification vs compression)
- λ= nonlinear saturation

Critical condition

$$\mu = 1$$

- $\mu < 1$: symmetry restored

- $\mu > 1$: stable asymmetry

Stable attractor

$$b_* = \sqrt{\frac{\mu - 1}{\lambda}}$$

6. Axiom VI — Compression vs Amplification

Two competing processes govern recursion:

Compression (universal scaling)

$$b \to \alpha_F^{-1} b (\alpha_F^{-1} \approx 0.4)$$

Amplification (selection)

$$b \to 2b \text{ (Born-like quadratic weighting)}$$

Stability condition

$$\text{amplification} \approx \text{compression}$$

This defines the emergence threshold.

7. Axiom VII — Collapse as Constraint Crossing

Collapse is not probabilistic, but structural:

$$S(\Psi) \geq R_{\text{system}} \Rightarrow \text{collapse}$$

Interpretation:

- entropy exceeds resolution capacity

- coherence fails
- system transitions to stable attractor

8. Axiom VIII — Time as Emergent

Time is not fundamental.

$$t \propto \varepsilon$$

Time is the ordering of recursive entropy resolution.

9. Axiom IX — Energy as Entropic Gradient

Energy is not primitive.

$$E \propto \nabla S$$

Energy is the momentum of entropy resolution.

10. Axiom X — Planck Constraint

Recursive resolution has a lower bound:

$$\tau \geq t_P$$

Where:

- t_P= Planck time

Below this, structure cannot form.

11. Axiom XI — Geometry from Recursion

Dimensionality emerges when resolution stabilizes:

$$D \propto \frac{R(\tau)}{S(\tau)}$$

- stable ratio → persistent structure
- unstable → collapse

12. Axiom XII — Gauge Structure as Recursion Hierarchy

- U(1): phase
- SU(2): bifurcation (symmetry breaking)
- SU(3): closure (confinement)

13. Axiom XIII: The Negative Feedback Principle

No recursively stabilising system can persist through unrestricted accumulation. As retained structure increases, destabilising pressure rises proportionally, producing oscillatory correction and preventing terminal recursive collapse.

$$\frac{d\Upsilon}{d\varepsilon} = \alpha - \beta\Upsilon + \gamma \sin{(\omega\varepsilon)}$$

where:

- α represents stabilising accumulation,
- $\beta\Upsilon$ represents rigidity-dependent negative feedback,
- $\gamma \sin(\omega\varepsilon)$ represents oscillatory entropy reinjection.

Key relation

The cubic term b^3 represents the minimal recursive signature of closure. Quadratic systems permit opposition and oscillation, but cubic structure permits stable triadic containment. In this sense, the emergence of third-order recursive closure corresponds naturally to the appearance of

SU(3), the minimal non-Abelian symmetry supporting persistent confined structure.

14. Core Principle

Structure exists only where recursive entropy resolution outpaces entropy accumulation

15. Derived Consequences

From the above:

- Born rule → minimal stability weighting
- collapse → threshold crossing
- mass → stabilized recursive curvature
- time → ordered resolution
- gravity → entropy gradient descent
- cognition → entropy filtering
- disease → entropy > resolution

16. Minimal Statement of DE

Deductive Entropy describes reality as the recursive resolution of uncertainty under constraint, where structure emerges only when entropy can be compressed faster than it accumulates, and persists only when memory stabilizes that compression across iterations.

Appendix E: Topological and Representation Structure

E.1 Overview

This appendix collects the mathematical structures underlying the Deductive Entropy (DE) framework. The main text uses these results at a conceptual level. Here we state them explicitly, where the central idea is that topology encodes memory, symmetry encodes admissible transformations, and their interaction defines the space of persistent structures.

E.2 Topology as Memory: Riemann Surfaces and Genus

Let Σ_g be a compact oriented Riemann surface of genus g. Its fundamental group is:

$$\pi_1(\Sigma_g) = \langle a_1, b_1, \dots, a_g, b_g \mid \prod_{i=1}^{g}[a_i, b_i] = 1\rangle$$

Each pair (a_i, b_i) corresponds to a non-contractible cycle. Each cycle supports holonomy as path-dependent transformations, so that non-contractible cycles cannot be removed by local deformation. Therefore genus g determines the number of independent memory channels;

- $g = 0$: no persistent memory
- $g \geq 1$: intrinsic, non-erasable memory

E.3 SU(n) as State-Transition Algebra

Let $G = \mathrm{SU}(n)$. A flat connection on Σ_g corresponds to a homomorphism:

$$\rho: \pi_1(\Sigma_g) \to \mathrm{SU}(n)$$

modulo conjugation. The space of such representations is:

$$\mathrm{Hom}(\pi_1(\Sigma_g), \mathrm{SU}(n))/\mathrm{SU}(n)$$

Dimension scaling

For irreducible representations:

$$\dim \mathcal{M}_{g,n} \sim (2g - 2)(n^2 - 1)$$

- $n^2 - 1$: internal degrees of freedom (algebra dimension)
- $2g - 2$: topological amplification

Thus genus sets capacity, SU(n) sets expressivity

E.4 Global Constraint and Coherence

The relation:

$$\prod_{i=1}^{g} [a_i, b_i] = 1$$

imposes a global constraint;

- memory is not independent per cycle
- states are globally coupled
- admissible configurations must satisfy a consistency condition

Thus information is relational, not additive.

E.5 Moduli Space as Information Manifold

The moduli space of flat connections:

$$\mathcal{M}_{g,n}$$

represents all globally consistent configurations;

- each point = a global memory state
- dimension = total information capacity
- structure = constrained, not combinatorial

Thus the moduli space is the phase space of structured memory.

E.6 Non-Abelian Memory and Irreversibility

For non-abelian groups (e.g. SU(n)):

$$UV \neq VU$$

- order of operations matters
- path-dependence is intrinsic
- sequence cannot be erased

This gives us an intrinsic arrow without external time.

E.7 Quantization and Compression

In quantum theory, admissible states are typically represented within a finite-dimensional Hilbert space. In topological formulations, such as Chern–Simons theory and conformal field theory, the dimensionality of this space is determined by global constraints, including the genus g as encoded by the Verlinde formula. From a DE perspective, this structure admits a more general interpretation. The transition from classical to quantum description can be understood as a reduction in accessible configurations under constraint. Classical configurations span a large, often continuous space, whereas quantum states occupy a smaller, discretised set of admissible configurations. This reflects three consistent features:

- classical configurations are compressed into a reduced set of stable states
- information is not lost, but reorganised into a more constrained representation
- resolution decreases while stability increases

Quantization can therefore be interpreted as a process of entropic compression under constraint, where only configurations capable of persisting under recursive stability are retained. In this framework, Hilbert space represents a derived representation of the admissible state space after compression, which provides an efficient mathematical description of stable configurations, but the underlying selection process is governed by recursive constraint rather than imposed structure.

E.8 Spinors and Minimal Representation

Spinors arise as minimal representations of orientation and phase. A null vector can be written:

$$k^{AA'} = \xi^A \bar{\xi}^{A'}$$

- ξ^A: carries phase and orientation
- complex structure encodes helicity
- minimal object for null structure

Thus spinors are minimal carriers of directional information.

E.9 Twistor and Bi-Twistor Structure

A twistor $Z^{\alpha} = (\omega^{A}, \pi_{A'})$ satisfies the incidence relation
$\omega^{A} = ix^{AA'}\pi_{A'}$

A bi-twistor $X^{\alpha\beta} = Z^{[\alpha}W^{\beta]}$ encodes relations between states.

- twistor: single direction + phase
- bi-twistor: relation between directions
- separation encoded algebraically

Thus geometry emerges from relational structure.

E.10 Mapping to DE Variables

The DE axes map naturally onto this structure:

Mathematical object	DE interpretation
holonomy	memory
genus g	memory depth
SU(n)	transformation algebra
moduli space	information manifold
spinor	phase + orientation
twistor	relational geometry
bi-twistor	entropic distance ε

E.11 Colour and Flavour as Memory Structures

Within this framework:

Colour (SU(3))

- closed holonomy loops
- exact cancellation constraint
- no external leakage

→ memory under closure

Flavour (weak sector)

- transitions between inequivalent states
- non-commuting transformations
- sequence-dependent

→ memory under update

- colour preserves internal consistency
- flavour updates the memory record

Thus strong interaction stabilises memory, weak interaction modifies it.

E.12 Synthesis

The combined structure can be summarised as:

- topology defines memory capacity
- symmetry defines transformation rules
- global constraints enforce coherence
- non-commutativity produces ordering
- quantization compresses representation

Persistent structure arises when memory, transformation, and constraint align. Geometry is the projection of this alignment, and matter is its stable realisation.

Appendix F: Technical Note on the $\boldsymbol{\varepsilon} - \boldsymbol{\Upsilon}$ Conjugate

This book has argued that reality can be understood as a balance between two fundamental tendencies. This section outlines a consistent mapping rather than a full derivation.

- the ability of a system to distinguish and resolve differences
- the ability of a system to retain structure and memory

We denote these as:

- **$\boldsymbol{\varepsilon}$ phase resolution**
- **$\boldsymbol{\Upsilon}$ committed structure**

The central claim is simple, that these two quantities are not independent. They are conjugate and as structure accumulates, flexibility is lost. As flexibility increases, structure dissolves. This trade-off is not just a feature of biology or cognition, it appears to operate at every scale. The purpose of this appendix is to show that, if this conjugacy is taken seriously, then gravity, time dilation, black holes and even the direction of quantum gravity all emerge as consequences of that single constraint.

1. The $\boldsymbol{\varepsilon} - \boldsymbol{\Upsilon}$ Conjugate Pair

We begin with the core relation:

$$\Delta\varepsilon\,\Delta\Upsilon \gtrsim \kappa$$

This expresses an intrinsic trade:

- increasing structural commitment (Υ) reduces available phase resolution (ε)
- increasing phase resolution reduces persistent structure

At the field level:

$$\frac{\partial \varepsilon}{\partial \Upsilon} < 0$$

This is the generative rule.

2. Emergence of Geometry

We treat ε as a spatial field:

$$\varepsilon = \varepsilon(x)$$

and define the effective metric as a function of ε:

$$g_{00} \sim \varepsilon g_{rr} \sim \varepsilon^{-1}$$

So:

- ε sets local clock rate
- its reciprocal sets spatial scaling

Curvature appears wherever ε varies:

$$\text{curvature} \sim \nabla\nabla\varepsilon$$

Thus spacetime is not fundamental in this frame, it is the large-scale expression of how ε is distributed under structural load.

3. Gravity

Structural density Υ sources ε through:

$$\nabla^2 \varepsilon \propto -\rho$$

For a point mass:

$$\varepsilon(r) = \varepsilon_\infty - \frac{C}{r}$$

Motion follows paths that minimise recursive cost:

$$\mathcal{C} = \int \frac{ds}{\varepsilon(x)}$$

which yields:

$$\mathbf{a} \propto -\nabla\Phi(\varepsilon)$$

Recovering:

$$\mathbf{a} \sim -\frac{1}{r^2}$$

So gravity is thus motion along gradients in ε induced by Υ.

4. Schwarzschild Sector

Matching to the weak-field limit fixes:

$$\frac{\varepsilon(r)}{\varepsilon_\infty} = 1 - \frac{2GM}{rc^2}$$

Defining the metric:

$$ds^2 = -\varepsilon_* c^2 dt^2 + \varepsilon_*^{-1} dr^2 + r^2 d\Omega^2$$

yields the Schwarzschild solution;

- time dilation is reduced ε
- radial stretching is reciprocal compensation
- horizon is where $\varepsilon_* = 0$

5. Classical Tests

From this ε-metric:

Gravitational redshift

$$\frac{\nu_\infty}{\nu_r} = \sqrt{\varepsilon_*}$$

Light bending

$$\Delta\phi = \frac{4GM}{bc^2}$$

Perihelion precession

$$\Delta\phi = \frac{6\pi GM}{a(1 - e^2)c^2}$$

All arise from the same field:

$$\varepsilon_*(r) = 1 - \frac{2GM}{rc^2}$$

6. Rotation and the Kerr Analogue

Rotation introduces a second component:

- scalar ε depletion
- azimuthal ε-current

The metric acquires a cross-term:

$$g_{t\phi} \neq 0$$

giving frame dragging:

$$\omega_{\text{drag}} \sim \frac{GJ}{c^2 r^3}$$

- rotation = circulating phase bias
- frame dragging = advection of local recursion frames
- ergosphere = no stationary frame possible
- horizon = no outward recursive closure

7. Time Dilation Unified

Proper time scales with ε:

$$d\tau \propto \varepsilon \, d\lambda$$

Two mechanisms reduce ε:

- structural load (gravity)
- directional coherence (velocity)

Hence:

$$d\tau = dt\sqrt{1 - \frac{2GM}{rc^2}} d\tau = dt\sqrt{1 - \frac{v^2}{c^2}}$$

Both arise from the same constraint.

8. Toward Quantum Gravity

If spacetime is emergent, it should not be quantized directly. Instead, quantize the conjugate pair:

$$[\hat{\varepsilon}(x), \hat{\Upsilon}(y)] = i\kappa\, \delta(x - y)$$

Then:

- geometry = coarse-grained ε field
- curvature = redistribution under Υ
- gravitons = excitations of emergent geometry

The Planck scale marks breakdown of the smooth $\varepsilon \rightarrow$ metric mapping, not the breakdown of reality.

Conclusion

The $\varepsilon - \Upsilon$ conjugate reframes physics around a single principle:

reality is governed by a constrained trade between phase resolution and committed structure.

From this:

- gravity is the redistribution of ε
- time dilation is reduced recursive capacity
- curvature is the geometry of ε variation
- black holes mark limits of accessible ε
- quantum gravity lies beneath spacetime, not within it

What begins as a constraint between phase resolution and structural commitment unfolds across every scale at which systems persist. In this light, gravity is no longer an external force, time is no longer a primitive backdrop, and spacetime itself is no longer fundamental. Each is a macroscopic expression of how reality allocates its capacity to distinguish and to endure. The same constraint that governs matter and motion also governs perception and thought. What appears as physics at one scale and experience at another is, in this sense, the same structure seen from different sides. The task that follows is to recognise and integrate our existing theories as limits of a deeper, shared framework.

Appendix G: Oscillatory Correction, Negative Feedback Dynamics, and the Emergence of SU(2)

A central implication of Deductive Entropy (DE) is that recursively stabilising systems cannot persist through monotonic accumulation alone. Any system that continuously compresses uncertainty into retained structure without compensatory destabilisation eventually undergoes recursive closure and collapse into rigidity.

This appendix formalises the role of oscillatory correction and demonstrates how dyadic symmetry naturally emerges once recursive systems require bounded self-correction (Chapter 13: Origin). The treatment presented here incorporates conceptual discussions and derivational insights arising from ongoing collaboration and personal communication with Sidney Belzberg.

1. Recursive Accumulation and Structural Instability

Within DE:

- εrepresents entropic distance,
- Υrepresents retained stabilised structure or committed memory.

A purely accumulative recursive system may be written:

$$\frac{d\Upsilon}{d\varepsilon} = \alpha$$

where:

- α represents stabilising accumulation.

This system exhibits monotonic growth:

$$\Upsilon(\varepsilon) = \alpha\varepsilon + \Upsilon_0$$

Initially:

- coherence increases,
- structure stabilises,
- predictability improves.

However no compensatory mechanism exists to oppose continued structural retention.

As:

$$\Upsilon \to \infty$$

the system progressively loses adaptive flexibility:

- perturbation recovery weakens,
- contradiction suppression increases,
- recursive closure emerges.

This represents a DE analogue of singularity formation.

A2. Negative Feedback Dynamics

To avoid runaway recursive accumulation, stabilising systems require negative feedback. Introducing proportional destabilisation yields:

$$\frac{d\Upsilon}{d\varepsilon} = \alpha - \beta\Upsilon$$

where:

- $\beta\Upsilon$ represents destabilising pressure proportional to accumulated rigidity.

The solution becomes bounded:

$$\Upsilon(\varepsilon) = \frac{\alpha}{\beta} + Ce^{-\beta\varepsilon}$$

where:

- C is an integration constant determined by initial conditions.

The system now approaches dynamic stability rather than runaway accumulation. Importantly:

- increasing retained structure generates increasing destabilising pressure,
- rigidity itself becomes the source of corrective force.

Negative feedback therefore prevents irreversible recursive closure.

3. Oscillatory Entropy Reinjection

Bounded accumulation alone is insufficient to explain adaptive persistence. Living systems, cognitive systems, ecosystems, and cosmological structures exhibit periodic destabilisation and renewal. To represent this behaviour, DE introduces oscillatory entropy reinjection:

$$\frac{d\Upsilon}{d\varepsilon} = \alpha - \beta\Upsilon + \gamma\sin(\omega\varepsilon)$$

where:

- $\gamma\sin(\omega\varepsilon)$ represents oscillatory destabilisation,
- ω represents recursive oscillatory frequency.

The system now alternates between:

- stabilisation,

- release,
- reorganisation,
- renewed stabilisation.

Oscillation therefore emerges naturally from recursive systems containing:

- stabilising accumulation,
- negative feedback correction,
- periodic entropy reinjection.

This produces bounded recursive survivability.

A4. The Necessity of Dyadic Structure

A scalar recursive state can accumulate or decay, but it cannot internally encode reversible oscillatory self-correction while preserving phase coherence. Oscillatory correction requires paired conjugate structure. DE therefore introduces the minimal recursive phase object:

$$\Psi = \begin{pmatrix} \psi_+ \\ \psi_- \end{pmatrix}$$

where:

- ψ_+ represents stabilising accumulation,
- ψ_- represents destabilising release or entropy reinjection.

The recursive state is now fundamentally dyadic. This transition is unavoidable once reality must preserve:

- bounded oscillation,
- reversible correction,

- recursive phase continuity.

The emergence of two-component recursive structure therefore follows directly from the requirement that adaptive systems avoid terminal rigidity.

A5. Norm Preservation and the Emergence of SU(2)

To preserve recursive stability, transformations acting on the dyadic state must preserve total recursive amplitude. The transformed state is written:

$$\Psi' = U\Psi$$

with norm preservation requiring:

$$U^{\dagger}U = I$$

where:

- $U^{\dagger}$ is the Hermitian conjugate of U,
- I is the identity operator.

To preserve orientation and exclude trivial global scaling:

$$\det(U) = 1$$

Thus the allowed transformations satisfy:

$$U \in SU(2)$$

The special unitary group therefore emerges naturally as the minimal symmetry preserving bounded recursive oscillation under negative feedback dynamics. In DE:

$$SU(2) = \text{the symmetry of oscillatory self-correction}$$

This interpretation differs from viewing SU(2) purely as an abstract internal symmetry. Here it arises as a necessary consequence of recursive survivability.

A6. Critical Asymmetry and Recursive Memory Emergence

Oscillation alone is insufficient to generate persistent reality. A perfectly symmetric oscillatory system continuously reverses its own stabilisation. Every retained recursive trajectory is ultimately cancelled by an opposing phase reversal. While bounded oscillation persists, no net structural memory survives. At exact recursive symmetry:

$$P_s = P_r = 0.5$$

where:

- P_s represents stabilising persistence probability,
- P_r represents recursive reversal probability.

Under this condition:

$$\Sigma(\varepsilon) \rightarrow 0$$

No committed structural asymmetry accumulates across recursive traversal. The system oscillates without remembering. DE therefore predicts that persistent reality requires a slight excess of stabilising retention over recursive cancellation. A minimal survivable asymmetry may be expressed conceptually as:

$$P_s \approx 0.5005 > P_r \approx 0.4995$$

This small imbalance produces profound long-term consequences. Across sufficiently large recursive depth:

- stabilised phase relations survive slightly more frequently than they reverse,
- retained structure accumulates incrementally,

- recursive trajectories become path-dependent,
- committed memory emerges.

The universe now carries retained structural history forward through recursive traversal:

$$\Sigma(\varepsilon) = \int_0^{\varepsilon} \Upsilon(\varepsilon')\, d\varepsilon' > 0$$

This transition marks the emergence of physical memory within the DE framework. Reality becomes directional because recursive stabilisation is no longer perfectly reversible.

A7. Oscillatory Persistence and the Origin of Time

Once recursive memory accumulates, future states become conditioned by prior stabilisation. The universe no longer exists as isolated transient oscillations, as it develops committed recursive history. This introduces temporal asymmetry, and time therefore emerges as the directional accumulation of retained recursive structure across entropic traversal. The emergence sequence becomes:

Stage	DE Process
Symmetric fluctuation	transient reversible oscillation
Negative feedback emergence	bounded recursive stability
Oscillatory correction	adaptive survivability
Critical asymmetry	persistent phase retention

Stage	DE Process
Recursive memory accumulation	committed structural history
Temporal emergence	directional recursive persistence

Within this framework:

- oscillation prevents recursive singularity,
- asymmetry permits retained memory,
- memory generates temporal directionality.

The universe therefore evolves dynamically as it recursively remembers its own stabilisation history.

A8. Ontological Consequence

The harmonic oscillator appears ubiquitously throughout physics because oscillatory correction is among the simplest mechanisms capable of preserving bounded recursive survivability. DE predicts that adaptive systems persist through the balance between:

- stabilisation and release,
- memory and uncertainty,
- coherence and destabilisation.

A universe without oscillatory correction collapses into rigidity. A universe without slight stabilisation asymmetry cannot remember. Existence therefore emerges from the coupled interaction of:

- bounded oscillation,

- negative feedback regulation,
- and persistent recursive asymmetry.

In this interpretation, the universe survives because it oscillates. The universe persists because it remembers.

Appendix H: Recursive Horizons, Chiral Recursion, and the Penrose Geometry of Memory

In General Relativity, the Schwarzschild horizon is usually treated as a causal boundary in spacetime. In Schwarzschild coordinates, the metric appears singular at the radius

$$r_s = \frac{2GM}{c^2}$$

where:

- r_s= Schwarzschild radius
- G= gravitational constant
- M= gravitating mass
- c= speed of light

However, this singularity disappears under coordinate transformation, implying that the horizon itself is not physically pathological. In the Deductive Entropy (DE) framework, this observation becomes foundational. The horizon is not interpreted as a coordinate singularity, but as a recursive fixed surface in the geometry of entropy resolution.

Null Expansion and Recursive Saturation

The invariant quantity defining the horizon is the outgoing null expansion:

$$\theta_{(k)}$$

where:

- $\theta_{(k)}$= expansion scalar of the outgoing null congruence

- k^a= null generator vector field tangent to the horizon

The Schwarzschild horizon is then defined invariantly by:

$$\theta_{(k)} = 0$$

Outside the horizon:

$$\theta_{(k)} > 0$$

meaning outgoing null rays diverge and recursive emergence proceeds normally. Inside the trapped region:

$$\theta_{(k)} < 0$$

meaning all future-directed null trajectories converge inward. In DE, recursive structural accumulation is represented by:

$$\Upsilon$$

where:

- Υ= accumulated structural memory or retained recursive constraint
- ε= recursion depth or entropic ordering coordinate
- α= local entropy-to-structure conversion coefficient

The recursive evolution law is:

$$\boxed{\frac{d\Upsilon}{d\varepsilon} = \alpha\theta_{(k)}}$$

Thus:

- outside the horizon:

$$\frac{d\Upsilon}{d\varepsilon} > 0$$

structure accumulates;

- at the horizon:

$$\boxed{\theta_{(k)} = 0 \Leftrightarrow \frac{d\Upsilon}{d\varepsilon} = 0}$$

recursive emergence reaches a fixed point;

- inside the trapped region:

$$\frac{d\Upsilon}{d\varepsilon} < 0$$

recursive orientation reverses.

The Schwarzschild horizon therefore becomes the invariant DE transition surface separating positive and negative recursive orientation.

Horizons as Chiral Pivots

The horizon carries an intrinsic orientation encoded by the horizon binormal:

$$\varepsilon_{ab}$$

where:

- ε_{ab}= antisymmetric horizon binormal tensor defining orientation of the horizon surface

In General Relativity, reversing orientation:

$$\varepsilon_{ab} \rightarrow -\varepsilon_{ab}$$

is merely a sign convention. In DE, however, this reversal corresponds to a genuine chiral transition in recursive orientation. Define the chiral recursion parameter:

$$\chi = \text{sgn}\,(\varepsilon_{ab})$$

where:

- χ= chiral recursion orientation parameter

Thus:

$$\chi_+ = +1, \chi_- = -1$$

The horizon separates two conjugate recursive orientations:

- positive recursive emergence,
- negative recursive emergence.

The horizon therefore acts as a chiral recursion pivot.

Horizon Entropy and Memory Preservation

The invariant entropy of a stationary horizon is the Wald entropy:

$$S_{\mathcal{H}} = -2\pi \int_{\Sigma} \frac{\partial \mathcal{L}}{\partial R_{abcd}} \varepsilon_{ab}\varepsilon_{cd}\sqrt{h}\, d^2x$$

where:

- $S_{\mathcal{H}}$= horizon entropy
- Σ= horizon two-surface
- $\mathcal{L}$= gravitational Lagrangian density
- R_{abcd}= Riemann curvature tensor
- h= determinant of induced horizon metric

In General Relativity this reduces to:

$$S_{\mathcal{H}} = \frac{k_B A}{4\ell_P^2}$$

where:

- k_B= Boltzmann constant
- A= horizon area
- ℓ_P= Planck length

Because entropy depends quadratically on the binormal:

$$\varepsilon_{ab}\varepsilon_{cd}$$

orientation reversal leaves entropy invariant:

$$\boxed{S_{\mathcal{H}}^{(+)} = S_{\mathcal{H}}^{(-)}}$$

However, the direction of entropy flow reverses:

$$\boxed{\frac{dS_+}{d\varepsilon} = -\frac{dS_-}{d\varepsilon}}$$

Thus:

- entropy magnitude is preserved,
- recursive orientation reverses.

The horizon therefore preserves information while reversing recursion direction.

Penrose Diagrams as Recursive Accessibility Maps

In standard General Relativity, Penrose diagrams compactify spacetime while preserving null and causal structure. The maximally extended Schwarzschild solution contains:

- Region I: external universe,
- Region II: black-hole interior,
- Region III: mirror asymptotic region,

- Region IV: white-hole region.

In DE, these become recursion domains rather than merely causal sectors.

Region I

Positive recursive emergence:

$$\frac{d\Upsilon}{d\varepsilon} > 0$$

Memory accumulates and entropy resolves into stable structure.

Horizon

The horizon satisfies:

$$\theta_{(k)} = 0$$

It is therefore:

- a recursive fixed surface,
- an entropy-preserving transition layer,
- a chiral orientation boundary.

Region II

Inside the trapped region:

$$\theta_{(k)} < 0$$

Recursive emergence becomes inward-directed and structure approaches recursive saturation. Thus the classical singularity is reinterpreted as recursive compression limit, with geometric encoding failure at maximal memory density.

Region III

Region III becomes the conjugate recursive branch:

- opposite entropy orientation,
- opposite chiral recursion direction,
- shared horizon geometry.

The Einstein–Rosen bridge is therefore interpreted as continuity of recursive topology across conjugate branches.

Raychaudhuri Equation and Recursive Compression

The invariant focusing equation for null congruences is the Raychaudhuri equation:

$$\frac{d\theta}{d\lambda} = -\frac{1}{2}\theta^2 - \sigma_{ab}\sigma^{ab} - R_{ab}k^a k^b$$

where:

- λ= affine parameter along null generators
- σ_{ab}= shear tensor of the null congruence
- R_{ab}= Ricci curvature tensor

At the horizon:

$$\theta = 0$$

so:

$$\frac{d\theta}{d\lambda}\Big|_{\mathcal{H}} = -\sigma_{ab}\sigma^{ab} - R_{ab}k^a k^b$$

In DE:

- $\sigma_{ab}\sigma^{ab}$ measures recursive shear distortion,
- $R_{ab}k^a k^b$ measures recursive curvature load.

The horizon therefore marks the transition where recursive expansion can no longer remain positive.

Recursive Horizon Functional

Define the DE horizon memory functional:

$$\Upsilon_{\mathcal{H}} = \int_{\Sigma} (\sigma_{ab}\sigma^{ab} + R_{ab}k^a k^b)\, dA$$

where:

- $\Upsilon_{\mathcal{H}}$= total recursive memory load stored at the horizon
- dA= differential horizon area element

This functional measures:

- retained recursive curvature,
- stored horizon memory,
- saturation of geometric encoding.

Black holes therefore become maximal recursive memory attractors.

Summary

In the DE framework:

- the Schwarzschild horizon is an invariant recursive fixed surface,
- horizons preserve entropy while reversing recursive orientation,
- Penrose diagrams become maps of recursive accessibility,
- singularities represent recursive saturation limits rather than physical infinities,

- and spacetime curvature emerges as the geometric bookkeeping of accumulated recursive memory.

Thus the horizon is the null surface at which recursive emergence stalls and memory becomes holographically encoded.

List of Variables and Definitions

Core DE Variables

ε (Entropic Coordinate / Resolution Depth)
Monotonic ordering variable indexing recursive entropy resolution. Replaces time as the fundamental progression axis.

Υ (Structural Commitment / Memory Depth)

Accumulated, stabilised constraint. Measures retained structure across recursion. Functions as a scalar complexity measure in descriptive contexts and as a generator operator — driving gauge rotations via $R(\varepsilon) = exp(i\varepsilon\Upsilon)$ — in the formal derivations. Also referred to as "complexity axis" or "entanglement depth" in the first book; "structural commitment" is the more precise term used throughout this volume.

$$\Sigma = \int d\Upsilon$$

Total retained phase corresponding to committed structure.

S (Entropy)
Measure of unresolved configurational freedom:

$$S = -\sum p_i log\ p_i$$

$R(\varepsilon)$ (Resolution Rate)

$$R(\varepsilon) = -\frac{dS}{d\varepsilon}$$

Rate at which entropy is resolved into structure.

Information and Prediction Variables

δ (Surprisal / Prediction Error)

$$\delta = -log\ P(y \mid \hat{y})$$

Information mismatch between prediction and observation.

$m(\varepsilon)$ (Internal Model)

System's compressed representation at recursion depth ε.

$\hat{y}(\varepsilon)$ (Prediction)

$$\hat{y}(\varepsilon) = P(m(\varepsilon))$$

Predicted observable derived from internal model.

$y(\varepsilon)$ (Observed Structure)

Realised structure in the environment.

Memory and Stability Variables

$M[\psi]$ (Memory Functional)

$$M[\psi] = \int \| \nabla\psi(x) \|^2\ dx$$

Cost of inconsistency in stored memory.

ρ_M (Memory Density)

Local density of retained structure.

I_M (Memory Persistence / Mutual Information)

$$I_M = I(X_\varepsilon; X_{\varepsilon+1})$$

Information retained across recursion.

Scaling and Renormalisation Variables

α (Feigenbaum Contraction Constant)

$$\alpha \approx 2.5029$$

Controls contraction of degrees of freedom across recursion.

δ (Feigenbaum Scaling Constant)

$$\delta \approx 4.669$$

Controls redistribution and concentration of retained structure.

N (Recursion Depth / Number of Scaling Steps)
Number of effective renormalisation levels.

λ (Ordering Parameter)
Alternative continuous recursion parameter.

Phase and Complex Variables

θ (Phase)
Relative ordering coordinate determining interference and memory.

ψ (Complex State Variable)

$$\psi = x + iy$$

Encodes magnitude and phase of structure.

φ (Phase Field)
Internal U(1) phase coordinate on memory fibre.

Gauge and Field Variables

A_μ (Gauge Connection / Memory Transport Field)
Defines how memory is transported across space.

$F_{\mu\nu}$ (Field Curvature / Memory Mismatch)

$$F_{\mu\nu} = \partial_\mu A_\nu - \partial_\nu A_\mu$$

Irreducible loop inconsistency.

$T_{\mu\nu}$ (Effective Stress–Energy Tensor)
Encodes total constraint contribution to curvature.

$g_{\mu\nu}$ (Metric Tensor)
Geometric representation of accumulated constraint.

Probability and Quantum Variables

ψ(x) (Amplitude)
Complex probability amplitude.

P(x) (Probability)

$$P(x) = |\psi(x)|^2$$

Probability of observing configuration x.

H (Hamiltonian)
Effective operator governing recursion dynamics.

Thermodynamic Variables

β (Marginal Cost of Memory)
Energy cost of adding structural commitment:

$$\beta = \frac{dE}{dY}$$

E (Energy)
Rate of entropy resolution.

t (Emergent Time)

$$t = \int \beta \, d\Upsilon$$

Accumulated cost of structure formation.

Collapse and Viability Variables

C_{delay} (Cost of Maintaining Uncertainty)
Accumulated cost of unresolved entropy.

C_{commit} (Cost of Commitment)
Cost of stabilising structure.

D (Attractor Depth)
Stability of attractor under perturbation.

Λ (Disturbance Load)
External/internal stress acting on system.

σ (Bias / Alignment Parameter)
Degree to which memory selects a specific attractor.

Geometry and Gravity Variables

Φ (Constraint Potential)

$$\Phi \sim \text{memory density}$$

Effective potential generated by accumulated constraint.

g (Acceleration / Gradient of Constraint)

$$g = -\nabla\Phi$$

r (Radial Coordinate)

Spatial coordinate in gravitational systems.

r_s (Schwarzschild Radius)

$$r_s = \frac{2GM}{c^2}$$

Cosmological Variables

Λ (Cosmological Constant)

Global bound on memory accumulation and curvature.

$\varepsilon \pm$ (Chiral Branches)

Dual orientations of emergence.

H (Hubble Parameter)

$$H = \frac{\dot{a}}{a}$$

Expansion rate of universe.

Biological and System Variables

B (Biomarker Burden / Constraint Source)

Source of biological constraint.

τ (Biological Time / Viability Horizon)

Remaining accessible future states.

X (System State Vector)

$$X = (x_1, x_2, \ldots, x_n)$$

Complexity and Correlation Variables

DoF (Degrees of Freedom)

Number of independent system states.

C (Complexity)

$$C = f(\text{DoF,constraint,memory})$$

TC (Total Correlation)

$$TC = \sum H(x_i) - H(X)$$

C_{eff} (Effective Dimensionality)

$$C_{\text{eff}} = \frac{(\sum \lambda_i)^2}{\sum \lambda_i^2}$$

Topology and Representation Variables

g (Genus)

Number of non-contractible cycles (memory channels).

π (Fundamental Phase Constant)

Defines rotational closure and chirality.

Z_3 (Centre Group of SU(3))

Three-state phase symmetry.

Variable	Meaning
ε	Ordering / resolution
ϒ	Memory / structure
Σ	Accumulated phase
S	Entropy
α	Contraction (how much survives)
δ	Redistribution (how structure reorganises)
β	Cost of memory
t	Emergent time
Φ	Constraint potential
Λ	Global constraint limit

Selected References

1. Penrose R. The Road to Reality. Jonathan Cape; 2004.
2. Penrose R. The Emperor's New Mind. Oxford University Press; 1989.
3. Penrose R. Shadows of the Mind. Oxford University Press; 1994.
4. Friston K. The free-energy principle: a unified brain theory? Nature Reviews Neuroscience. 2010;11(2):127 138.
5. Kauffman SA. The Origins of Order. Oxford University Press; 1993.
6. Kauffman SA. Investigations. Oxford University Press; 2000.
7. Prigogine I, Stengers I. Order Out of Chaos. Bantam Books; 1984.
8. Shannon CE. A mathematical theory of communication. Bell System Technical Journal. 1948;27:379–423, 623–656.
9. Gödel K. On Formally Undecidable Propositions of Principia Mathematica and Related Systems. Dover; 1992.
10. Einstein A. Relativity: The Special and the General Theory. Henry Holt; 1920.
11. Dirac PAM. The Principles of Quantum Mechanics. Oxford University Press; 1930.

12. Wheeler JA, Zurek WH, editors. Quantum Theory and Measurement. Princeton University Press; 1983.

13. Mandelbrot B. The Fractal Geometry of Nature. W.H. Freeman; 1982.

14. Feigenbaum MJ. Quantitative universality for a class of nonlinear transformations. Journal of Statistical Physics. 1978;19(1):25–52.

15. Hawking SW, Ellis GFR. The Large Scale Structure of Space-Time. Cambridge University Press; 1973.

16. Bohm D. Wholeness and the Implicate Order. Routledge; 1980.

17. Tegmark M. Consciousness as a state of matter. Chaos, Solitons & Fractals. 2015;76:238–270.

18. Tononi G. An information integration theory of consciousness. BMC Neuroscience. 2004;5:42.

19. Verlinde E. On the origin of gravity and the laws of Newton. Journal of High Energy Physics. 2011;2011(4):29.

20. Ioannidis JPA. Why most published research findings are false. PLoS Medicine. 2005;2(8):e124.